MUNICIPAL SUPPLIES

WATERWORKS — STREET — SEWERS
TRAFFIC SIGNALS — POLICE
FIRE EQUIPMENT

City Government Mural in Gloucester, Mass. City Hall. By Charles Allan Winter

W.S. Darley & Co.
World's Largest Makers of Municipal Supplies

CHICAGO
U. S. A.
Buyers Guide
No. 119

The Four Corners of *Your* Town

Do you havE this traffic confusion resulting in delays and accidents that can be avoided bv installing this highlv efficient and dependable traffic Signal

Our Sales prove this Signal to be the most popular in America

SIMPLEX TRAFFIC SIGNAL

Our Simplex signals operate for less. Savings amount to $50.00 to $95.00 a year, per signal. Not a statement but a FACT. In current saving alone THEY SAVE THE COST IN ONE YEAR over any other signal made. Something to consider these days, if you are trying to cut your budgets.

At 5c per kilowatt hours our 3-Color Signal operates for only 1½c per HOUR including motor controller.

Get the figures on the operating cost of OTHER makes and you will know this SIMPLEX will save you $50.00 to $95.00 a YEAR on current alone.

The scientific reason for Simplex efficiency is based on the law that any light throws its rays, illumination, or candle-power, EQUALLY IN ALL DIRECTIONS. Simplex Signals use 150-watt bulbs which are just as bright on one side as on the other, and where other signals have about 42 candle-power of light (Edison Co., Chicago, rating) Simplex Signals have 105 candle-power. Yet where other signals use 12 bulbs, Simplex Signals use but 3.

CLEAN CUT ENGINEERING SIMPLICITY

Interior View—with visors and lenses taken off—showing how superior, scientific design makes three bulbs do the work of twelve bulbs —— and do it much more efficiently.

Unequalled Value!

With a 5 Year Guarantee and 30 Days Free Trial.

All Traffic Guides and Controls made and sold by W. S. Darley & Co., are Warranted to be of the best materials, workmanship, and design and subject to the APPROVAL of the Buyers. If not satisfactory they are returnable, freight charges both ways paid by W. S. Darley & Co.

Any faulty, defective or unsatisfactory part will be replaced free of charge any time within 5 years from date of purchase, upon request by customer.

All shipments subject to 30 days' trial and approval.

W. S. DARLEY & CO.

Usual Selling Price $217.50 From W. S. Darley & Co—

$87⁵⁰
COMPLETE

**INCLUDING CONTROLLER
4 WAY CONSTRUCTION
WITH HIGH POWER LENS**

Rugged Construction

SIMPLEX Signals have housings or bodies made of heavy cast aluminum, the same as the costliest signals on the market. They are weatherproof and will NOT rust and NEVER corrode. Long sun visors, also aluminum, shadow the lenses even in direct sunlight. Even the light partition is aluminum. Hundreds of the Signals are in service throughout the country and will continue to give many years of loyal service, without interruption and virtually without ultimate wearout. They are a good safe investment and not an experiment.

On all four sides are convex high power lenses. They are set against felt washers and held in place by brass clips. The lenses proper are 3⅝" in diameter, the standard size. The colors will not fade and photometric tests prove these lenses are ideal for the requirements.

Any local electrician can easily hang and connect SIMPLEX Signals. Often it takes only a day, depending on how far to the power wire, etc. Wire and conduit cost very little.

Automatic—With Built-In Controller

NEW IMPROVEMENTS. Our Engineers designed a new Compact Motor Controller which is built into each signal. Heretofore Stop and Go lights had to have a SEPARATE CONTROLLER driven by a ¼ H. P. motor. Cost $42.75. We save you that sum because our price includes the controller—only 1/100 H. P. and it works perfectly. Also we save you wiring to the old style control. Just connect two wires to these SELF-CONTAINED signals and they begin to flash "STOP" and "GO."

No Sir! No Obligation to Buy of W. S. Darley & Co.

Trying Simplex Signals imposes no obligation of any kind. For they are sold only under our Free Trial policy. That is, we extend to every community the privilege of ordering Simplex Signals, of using them for a period of 30 days, and of then returning them if desired, at our expense for transportation both ways.

Many consider this center Suspended Type the equal, OR BETTER, than four corner signals on posts. That's because they are ALWAYS in full view and never out of sight by reason of cars ahead, or bus traffic, or partly obstructed by trees, poles, etc. Neither do they get knocked over or smashed. To try them is to like them.

Many Towns now operating two or four old style post signals at one corner, with 12 bulbs in each, could with ONE of our suspended signals, SAVE $95 to $285 a year on operating at 10c a kilowatt. Figure it out and see how the budget CAN BE REDUCED and money saved the taxpayers.

No. C811. Simplex Signal with RED, AMBER and GREEN lenses on all four sides, INCLUDING its SELF-CONTAINED MOTOR CONTROL, complete and all wired, ready to hang and operate **$87.50**

We set motor controls to flash 30-30 seconds with a 3 second amber light overlapping 1 second each on both red and green light, 30-20 seconds, 15-15 seconds, or any time intervals desired.

For a dended street intersection in a congested neighborhood requiring traffic control we can build a special C811 Simplex Traffic Signal. Same as C811 except that it is made 2 way, with one side blank. Operates same as C811 with 3 colors.

No. D407. Special Simplex Signal with Red, Amber and Green lenses on 3 sides only, including self-contained motor control, wired complete, ready to hang and operate **$79.50**

Read This!

**City of Roby
Texas**

W. S. Darley & Co.   Jan. 1, 1927.
Chicago, Ill.

Gentlemen:

We have been using one of your Simplex Traffic Signals for about two years and we find it one of the best investments our City has ever made.

Prior to the installation of this Traffic Light we were having an average of an accident about every three weeks and since the light has been installed we have not had an accident—BELIEVE IT OR NOT.

Yours very truly,
Geo. F. Britton, Mayor.

2

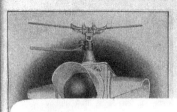

Simplex Traffic Signals
Designed and Built to Meet All State Laws

ONLY $99⁹⁵

From W. S. Darley & Co.—
Usual Selling Price $230.00

COMPLETE, INCLUDING BUILT-IN CONTROLLER
SEPARATE LUMINATION FOR EACH HIGH POWER LENS

Some States require separate lumination for each lens, and where this is a compulsory law we make the signal to comply with the law.

...ranged with all 4 Red at the top, 4 Amber in the ...and bulbs must not be less than 40 watts each. ...ur NATIONAL STANDARD SIGNAL. In ...1 Simplex Traffic Signal. Inside, however, ...t as other traffic signals, i. e., 12 reflectors and ...nd workmanship is all it should be in every ...$230 and get no more intrinsic value for your ...ould not offer to send it on strictly 30 DAYS ...low you to compare it with any other at any

AUTOMATIC CONTROLLER, set, as you may ...ROSS traffic for 30-30 seconds with a 3 second ...both red and green lights, 50-20 seconds, 15-15 ...o it's ready to OPERATE when hung and you ...troller but on wiring, etc. BELIEVE IT OR ...ke there isn't any traffic light so advanced in ...ours and with SELF-CONTAINED CONTROL...

. GIVING YOU DEPENDABLE A LONG AND ACTIVE LIFE

...lies made of heavy cast aluminum, the same as ...ey are weatherproof and will NOT rust and ...uminum, shadow the lenses even in direct sun...um. Hundreds of these Signals are in service ...e to give many years of loyal service, without ...ats wearout. They are a good safe investment

...ing and connect SIMPLEX Signals. Often it ...to the power wire, etc. Wire and conduit cost

National
Standard
Traffic
Signal

All National Standard Simplex Traffic Signals, both Suspended Type and Post Mounted Type, have a separate bulb behind each lens, as shown below.

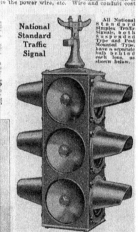

Interior View—with visors and lenses taken off—showing a separate bulb behind each lens. Conforms to laws in all states requiring this design.

Notice the reflectors, one for each bulb. Scientifically designed parabolic type, heavily and durably plated and highly polished.

SIMPLEX TRAFFIC SIGNALS ARE APPROVED by STATE of PA.

The Four Corners of *Your* Town

Do... ...ve this traffic confusion resultin... ...ccidents that can be... Highly efficient...

SIMPLEX TRA...

Our Simplex signals operate for less. ☛ Sav... amount to $50.00 to $95.00 a year, per signal. N... statement but a FACT. In current saving alone T... SAVE THE COST IN ONE YEAR over any other s... made. Something to consider these days, if yo... trying to cut your budgets.

☛ At 5c per kilowatt hours our 3-Color Sign... erates for only 1¼c per HOUR ☛ including... controller.

Get the figures on the operating cost of C... makes and you will know this SIMPLEX w... you $50.00 to $95.00 a YEAR on current alone.

The scientific reason for Simplex efficiency ... on the law that any light throws its rays, l... tion, or candle-power, EQUALLY IN ALL ... TIONS. Simplex Signals use 100-watt bul... are just as bright on one side as on the o... where other signals have about 43 candle... light (Edison Co., Chicago, rating) Simple... have 105 candle-power. Yet where other s... 12 bulbs, Simplex Signals use but 3.

CLEAN (
ENGINEE
SIMPLI(

Inter...
with v...
es tak...
ing ...
scien...
make...
do ...
twel...
do ...
eff...

Unequalled Value'
With a 5 Year Gua...
30 Days Free Trial.

All Traffic Guides and Controls made and sold by W. S. Darley & Co., are Warranted to be of the best materials, workmanship, and design and subject to the APPROVAL of the Buyers. If not satisfactory they are returnable, freight charges both ways paid by W. S. Darley & Co.

Any faulty, defective or unsatisfactory part will be replaced free of charge any time within 5 years from date of purchase, upon request by customer.

☛ All shipments subject to 30 days' trial and approval.

W. S. DARLEY & CO.

ntroller
SEPA-
hat sum
erfectly.
to these

sold only
rivilege of
n returning

four corner
out of sight
etc. Neither

corner, with
285 a year on
dget CAN BE

four
e and ... **$87.50**

mber light over-
seconds, or any

oper...
REDUCED ...
No. C811. Simplex ...
sides, INCLUDING its S...
all wired, ready to hang and ope...

☛ We set motor controls to flash 30-30 s... lapping 1 second each on both red and green light, o... time intervals desired.

☛ For a deadend street intersection in a congested neighborho... requiring traffic control ☛ we can build a special C811 Simplex Traffic Signal. Same as C811 except that it is made 3 way, with one side blank. Operates same as C811 with 3 colors. No. D497. Special Simplex Signal with Red, Amber and Green lenses on 3 sides only, including self-contained motor control, wired complete, ready to hang and operate **$79.50**

Simplex Traffic Signals

Designed and Built to Meet All State Laws

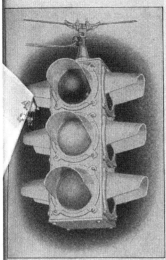

ONLY $99⁹⁵

From W. S. Darley & Co.—
☞ *Usual Selling Price $230.00*

COMPLETE, INCLUDING BUILT-IN CONTROLLER
SEPARATE LUMINATION FOR EACH HIGH POWER LENS

Some States require separate lumination for each lens, and where this is a compulsory law we make the signal to comply with the law.

Lenses, in this type, must be arranged with all 4 Red at the top, 4 Amber in the center and 4 Green at the bottom and bulbs must not be less than 40 watts each. This type and design we call our NATIONAL STANDARD SIGNAL. ☞ In appearance it is exactly like our C811 Simplex Traffic Signal. ☞ Inside, however, the construction is same arrangement as other traffic signals, i. e., 12 reflectors and 12 sockets for bulbs. The quality and workmanship is all it should be in every respect; you might pay as much as $230 and get no more intrinsic value for your money. If we did not think so we would not offer to send it on strictly 30 DAYS APPROVAL and FREE TRIAL to allow you to compare it with any other at any price.

INCLUDED! Inside is our marvelous AUTOMATIC CONTROLLER, set, as you may instruct us, to time THRU traffic and CROSS traffic for 30-30 seconds with a 3 second amber light overlapping 1 second each on both red and green lights, 20-20 seconds, 15-15 seconds, or any time intervals desired. So it's ready to OPERATE when hung and you not only save on buying a separate controller but on wiring, etc. BELIEVE IT OR NOT, as Ripley says, but to our knowledge there isn't any traffic light so advanced in engineering, so scientific, so compact as ours and with SELF-CONTAINED CONTROLLER for 3-color accurately timed signals.

ENDURINGLY RUGGED . . . GIVING YOU DEPENDABLE SERVICE THROUGHOUT A LONG AND ACTIVE LIFE

SIMPLEX Signals have housings or bodies made of heavy cast aluminum, the same as the costliest signals on the market. They are weatherproof and will NOT rust and NEVER corrode. Long sun visors, also aluminum, shadow the lenses even in direct sunlight. Even the light partition is aluminum. Hundreds of these Signals are in service throughout the country and will continue to give many years of loyal service, without interruption and virtually without ultimate wearout. They are a good safe investment and not an experiment.

☞ Any local electrician can easily hang and connect SIMPLEX Signals. Often it takes only a day, depending on how far to the power wire, etc. Wire and conduit cost very little.

☞ Automatic—With Built-In Controller

Our Engineers designed a new compact Motor Controller which is built into each signal, both Suspended Type and Post Mounted Type. Heretofore Stop and Go lights had to have a SEPARATE CONTROLLER driven by a ⅛ H. P. motor. Cost $42.75. We save you that sum ☞ because our price includes the controller—only 1/100 H. P. and it works perfectly. Also we save you wiring to the old style control.

Trying Simplex Signals impose no obligation of any kind. For they are sold only under our Free Trial policy. That is, we extend to every community the privilege of ordering Simplex Signals, of using them for a period of 30 days, and of then returning them, if desired, at our expense for transportation both ways.

Many consider this center Suspended Type the equal, OR BETTER, than four corner signals on posts. That's because they are ALWAYS in full view and never out of sight by reason of cars ahead, or bus traffic, or partly obstructed by trees, poles, etc. Neither do they get knocked over or smashed. ☞ To try them is to like them.

No. D290. National Standard Traffic Signal, Suspended Type, with Built-In Self-Contained Controller, all wired and ready to hang and connect.. **$99.95**

No. D290SA. National Standard Traffic Signal, Suspended Type, same as above but with Split Amber Controller so that amber light comes on ONLY when signal is changing from green to red. A safety feature that prevents waiting traffic from jumping the green light by starting when amber light shows. All wired and complete ready to hang and connect **$102.95**

W. S. Darley & Co. Offer ☞ 3% Discount for Cash

National Standard Post Mounted Type

Same construction as the No. D290 National Standard Suspended Type Traffic Signal, with separate lumination for each lens. There is a separate bulb behind each lens in this signal, as shown in interior view at right.

The Signal head is mounted on a steel post 4½" in diameter which is set in a heavy cast iron base that can be bolted to sidewalks or set on foundations in parkways.

For underground feed wiring comes up through the post. We also furnish it for overhead feed and wiring if the customer desires and specifies; no extra charge for overhead wiring connections. Any local electrician can make connections.

Automatic—With Built-In Controller

With a Controller inside each signal much hard labor and expense is saved tunneling under pavements to connect two or more corner signals to operate from a curb controller.

With our Simplex Control inside each signal TWO or FOUR or any number of signals will operate IN-STEP and properly show all colors at the right time and in the right way. It's very simple and as accurate as an electric clock. The control motors are SYNCHRONOUS and can be set to flash signals just as you want. ☞ Any number of signals along a street, or on corners, can be easily timed to show ALL GREEN at the same second, or timed PROGRESSIVELY. ☞ ALL WITHOUT ANY UNDERGROUND CABLES OR WIRING. Thru traffic and cross traffic can be timed different, as you may instruct us.

No. D220. National Standard Traffic Signal Post Mounted Type, with Built-In Self-Contained Controller, all wired and ready to connect............ **$99⁹⁵**

Customers furnish their own bulbs.
No. D220 does not include post. Steel Post 4" diameter with cast base .... **$14.00 EXTRA**

National Standard Traffic Signal

All National Standard Simplex Traffic Signals, both Suspended Type and Post Mounted Type, have a separate bulb behind each lens, as shown below.

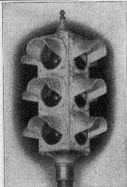

Interior View—with visors and lenses taken off—showing a separate bulb behind each lens. Conforms to laws in all states requiring this design.

☞ Notice the reflectors, one for each bulb. Scientifically designed parabolic type, heavily and durably plated and highly polished.

SIMPLEX TRAFFIC SIGNALS ARE APPROVED by STATE of PA.

3

A Signal That Leads the World in Value $67⁵⁰

SIMPLEX
extraordinary features plus
a Phenomenal Low Price

COMPLETE, INCLUDING
BUILT-IN CONTROLLER

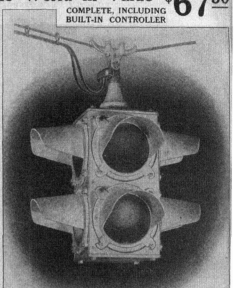

This new type meets the demand of many cities and villages for 2-Color signals. They are ideal; low in price, easily hung up, simply controlled, and very efficient.

The operation is quite simple. When the red STOP goes out, there is an interval of about three seconds before the green GO lights up, and vice versa. This interval gives the driver time to get ready.

The Lowest Cost of Operation

AT 10 CENTS PER KILOWATT HOUR THIS SIMPLEX SIGNAL ☞ INCLUDING ITS MOTOR CONTROLLER ☞ OPERATES FOR 2 CENTS PER HOUR.

The scientific reason for Simplex efficiency is based on the law that any light throws its rays, illumination, or candle-power, EQUALLY IN ALL DIRECTIONS. Simplex Signals use 150-watt bulbs which are just as bright on one side as on the other.

Where other signals of this type use 8 bulbs, this Simplex takes only 2 bulbs. Where other signals take 265 watts, this Simplex uses only 165 watts. They save $40 to $60 a year on current alone.

☞ And where 43 candlepower illuminates the lenses of other signals, ☞ 195 candlepower lights Simplex lenses.

NO maze of wires and sockets and "whatnots." Simplex Signals are NOT complicated.

No wonder Simplex Signals do not get out of order. No wonder repairs are never needed.

☞ Any local electrician can easily hang SIMPLEX Signals. Often it takes only a day, depending on how far to the power wire, etc. Wire and conduit cost very little.

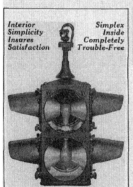

Interior Simplicity Insures Satisfaction

Simplex Inside Completely Trouble-Free

FOR CASH WITH ORDER TAKE OFF
3%
DISCOUNT

No maze of wires and sockets and "what-nots." Simplex Signals are NOT complicated.

No wonder Simplex Signals do not get out of order. No wonder repairs are never needed.

☞ Any local electrician can easily hang and connect SIMPLEX Signals. Often it takes only a day, depending on how far to the power wire, etc. Wire and conduit cost very little.

☞ Automatic—With Built-In Controller

Our Engineers designed a new Compact Motor Controller which is built into each signal. Heretofore Stop and Go lights had to have a SEPARATE CONTROLLER driven by a ⅛ H. P. motor. Cost $42.75. We save you that sum ☞ because our price includes the controller —only 1/100 H. P. and it works perfectly. Also we save you wiring to the old style control. ☞ Just connect two wires to these SELF CONTAINED signals and they begin to flash "STOP" and "GO."

Heavy Cast Aluminum Construction

SIMPLEX Signals have housings or bodies made of heavy cast aluminum, the same as, and equal in quality to the costliest signals on the market. They are weatherproof and will NOT rust and NEVER corrode. Long sun visors, also aluminum, shadow the lenses even in direct sunlight. Even the light partition is aluminum. Hundreds of these Signals are in service throughout the country and will continue to give many years of loyal service, without interruption and virtually without ultimate wearout. The construction of design is a good safe investment and not an experiment.

On all four sides are convex high power lens. They are set against felt washers and held in place by brass clips. The lenses proper are 8⅜" in diameter, the standard size. The colors will not fade and photometric tests prove these lenses are ideal for the requirements.

No. CS10. Simplex Signal with RED and GREEN lenses on all four sides, INCLUDING its SELF-CONTAINED MOTOR CONTROL, complete and all wired, ready to hang $67⁵⁰ and operate...............

Our Customers Sell Themselves
From Our Buyer's Guide

Its only traveling expenses are a postage stamp. It has no hotel bills; receives no salary or commissions to be added to the price of equipment you buy from us.

Read What This Official Has to Say About Simplex Traffic Signals

W. S. DARLEY & CO.
Chicago, Ill.
Gentlemen:

You will find attached check for $65.80 in payment of your invoice for CS10 Simplex Signal. We are well pleased with the Simplex Signal.

Yours very truly,
JAMES McDONALD, Mayor
Winfield, Alabama.

4

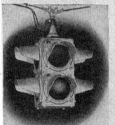

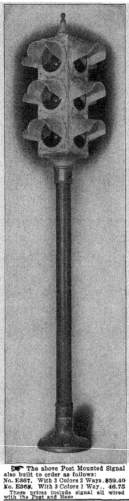

Simplex 3 Color 4 Way Post Mounted

The efficiency of this design of standard 3-Color 4-Way Signal makes it worthy of consideration. Strength and simplicity of construction reduce installation problems and expenses to a new minimum.

The body is cast aluminum—non-corrosive, rigid and durable. It is assembled weatherproof, with shadow-less and lightproof compartments. The long sun visors shadow the lenses even in direct sunlight. Instead of twelve compartments this Simplex Signal has only three divisions. And instead of twelve 60-watt lamps it has only three 100-watt lamps. More illumination and less expense. Easily accessible for cleaning or replacement of bulbs.

On each side there are three convex lenses of standard colors—Red, Amber and Green. They are set against felt washers and will not crack from vibration.

The Signal head is mounted on a steel post 8½" in diameter which is set on a heavy cast iron base that can be bolted to sidewalks or set on foundations in parkways.

The underground feed wiring comes up through the post. We also furnish it for overhead feed and wiring if the customer desires and specifies; no extra charge for overhead wiring connections. Any local electrician can make connections to the control box and current supply.

OPERATION EXPENSE

While this is a signal of superior brilliancy because it uses 100-watt bulbs, the cost of current is less than signals having twelve 60-watt bulbs. At 5c per kilowatt hour this Signal including its Motor Controller operates for 1½ cents per hour. Self contained units operate for 1½ cents per hour.

Use the figures on the operating cost of OTHER makes and you will know this SIMPLEX will save $50 to $80 a YEAR on current alone.

The prices of other competitive signals, of this same type, are very much higher because of complicated construction and multiplicity of parts which only add a heavy charge for bulbs, and nothing to the value.

No. C371. Three-color, four-way, post mounted Simplex Signal. All wired and ready to connect. Approximate shipping weight 300 lbs. Each.................$86.52

Also built with SELF CONTAINED CONTROL.

This is a new wonderful improvement! Instead of running power wires to a curb control box (like No. 715) and then underground cables to each of the color lights, we can put our new compact controller INSIDE THE SIGNAL so that only a single 2-strand wire is necessary. The wire may come up inside the post or down from OVERHEAD and through the top of the signal, as you prefer.

With a Controller inside each signal much hard labor and expense is saved tunneling under pavements to connect two or more corner signals to operate from a curb controller.

With our Control inside each signal TWO or FOUR or any number of signals will operate IN-STEP and properly show all colors at the right time and in the right way. It's very simple and as accurate as an electric clock. The control motors are SYNCHRONO's and can be set to flash signals just as you want. Any number of signals along a street, or on corners, can be easily timed to show ALL GREEN at the same second, or timed PROGRESSIVELY ALL WITHOUT ANY UNDERGROUND CABLES OR WIRING. Thru traffic and cross traffic can be timed different, as you may instruct us.

With self-contained Controllers these Simplex Signals, at 5c kilowatt, operate for 1½ cents per hour.

Specify on your order if the new D493 Automatic Simplex Controller, shown below, is wanted. We supply it installed complete in any of our Simplex Traffic Signals.

Simplex 2-Color 4-Way Post Mounted

This new type meets the demand of many cities and villages for 2-Color signals. They are ideal; low in price, easily set up, simply controlled, and very efficient.

The operation is quite simple. When the red STOP goes out there is an interval of about three seconds before the green GO lights up, and vice versa. This interval gives the driver time to get ready.

No. D496. Two-color, four-way, post mounted Simplex Signal. All wired and ready to connect. Each.................$69.50

SIMPLEX ADJUSTABLE SIGNALS

Suspended Type

☞ *The Best For Diagonal Street Intersections*

Each individual unit of this suspended type is adjustable to any exact angle to throw the beams of light where a diagonal street crosses a straight street. This type of Signal is evenly balanced. SIMPLEX construction throughout with our special long distance high power lens and durably plated and highly polished parabolic reflectors.

Adjustable to any angle, so the light beams will be just exactly right to all approaching vehicle drivers, to guide traffic without confusion.

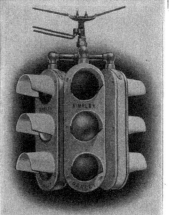

Including Automatic Controller

The Simplex Adjustable Suspended Signal is complete ready to install. Included is our special Simplex Controller in separate, cast aluminum, weatherproof box ready for installation at curb.

No. D489. Simplex Adjustable Suspended 3-Color 4-Way Signal, wired complete, including control...................$149.50

No. D490. Simplex Adjustable Suspended 3-Color 2-Way Signal, wired complete, including control...................$126.50

No. D491. Simplex Adjustable Suspended 2-Color 4-Way Signal, wired complete, including control...................$134.50

No. D492. Simplex Adjustable Suspended 2-Color 2-Way Signal, wired complete, including control...................$116.50

☞ The above Post Mounted Signal also built to order as follows:

No. E367. With 3 Colors 2 Ways..$59.40
No. E368. With 3 Colors 1 Way... 46.75
These prices include signal all wired with the Post and Base.

Aluminum Box for Traffic Controls

No. D498. Special Cast Aluminum Weather-proof Box for Simplex Automatic Controller, with special brackets for mounting on any post, only....$9.75

THE MARVELOUS AUTOMATIC SIMPLEX CONTROLLER

It would seem that skill could go no further in ingenuity. LOOK AT IT! Until Engineers for DARLEY developed this, a traffic controller was a bulky thing, consisting of a drum operated by a H. P. motor and selling for $47.50 or more.

This SIMPLEX controller contains a tiny motor of about 1/100 H. P. and operates your signal for about 1 cent per month. Old type controllers, ½ H. P., cost about 20 cents per 24 hours operation, or $73.00 a year as compared to only 12 cents a year for the Simplex. And it's so rugged it could carry ten times the load. No oiling! No attention! No brushes! It's a special motor as simple as the motor in Electric Clocks, without commutator or brushes. We'll guarantee it for 43,800 hours of continuous service, or five years.

This is the Control included with every Simplex and National Standard Traffic Signal from Darley & Co. Built-in, wired, ready to operate, IT'S SYNCHRONO too, so two or more signals can be set to step in time together, or vary as you may want them to work.

You can order the Automatic Controller for installation in any Traffic Signal, our own Simplex or any other make. Easily installed in Signals now in service. Or the Simplex Automatic Controller can be ordered complete in our special cast aluminum weatherproof box to set at curb.

No. D493. Automatic Simplex Controller, for any 3-Color Traffic Signal, only..$29.25

No. D494. Automatic Simplex Controller, for any 2-Color Traffic Signal, only..$28.25

No. D493X. Automatic Simplex Controller, same as No. D493, but with extra contact built in for when signal control, so that amber light comes on ONLY when signal is changing from green to red. A safety feature that prevents waiting traffic from jumping the green light before starting when amber light shows. For any 3-Color Traffic Signal, only..$34.25

☞ We set motor controls to flash 20 second traffic changes, but can supply to allow THRU traffic 30 seconds and CROSS traffic 20 seconds, without extra charge.

COMPLETE ONLY $29.25

8

Simplex Post Mounted Signals

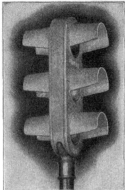

No. C264 Simplex Traffic Signal

Always in the direct line of vision before the eyes of drivers from both directions. With its great large 8⅝" special prismatic lens, it flashes a warning hundreds of yards, through storm or rain, to all approaching traffic.

Modernistic distinctive design. Built by the world's leading traffic light engineers. Every test and any comparison proves it far excels every other signal in its field.

SPECIAL CONSTRUCTION DETAILS

These SIMPLEX Traffic Signals have a cast aluminum housing of molded construction, fronts are removable and the body is a seamless, jointless, one-piece unit, which is an original feature by Darley pioneer designers.

Each lens has its own individual optical system, separated by section partitions also cast integral with the body.

Over each lens there is a cast aluminum visor which protects the signal from the direct rays of the sun. The front has gasket channels and hermetically seals all the compartments.

The complete Signal is aluminum, cast to our specifications in special molds.

PRISMATIC LENSES

The colored lens are mounted on felt dust-proof gaskets held in place by brass clips and can be wiped outside and inside without the removal of any parts. The lenses proper are 8⅝" in diameter, ¼" thick, single piece of glass with an outside smooth surface; the prismatic inside surface is provided with a prism construction. The lenses direct an intense beam of light for distant points on the street with a spread of light downward and outward to take care of nearby points. Photometric tests of the actual distribution of light prove these lenses are ideal for the requirements.

CHROMIUM PARABOLIC REFLECTORS

Each of the colored lenses is backed by an individual parabolic reflector, adopted specially for SIMPLEX Signals. These are 8¼" in diameter, CHROMIUM plated and polished on brass. When the Signal is opened they can be wiped clean with a dry cloth. Long life and maximum efficiency is assured because CHROMIUM will not tarnish, chip or break and the reflecting surface exceeds by nearly 100% the usual allowance.

Any standard 60 watt lamp can be used. Lamps are horizontally mounted and screwed into bases mounted on elastic shock-absorbers, to eliminate vibration and prolong the burning hours of the lamp.

WROUGHT STEEL POSTS

This type of SIMPLEX Signal is mounted on a wrought steel post 3½" outside diameter set in heavy flanged cast bases. The steel post easily supports the Signal rigidly, and will not chip or crack. Steel posts are frost-proof and practically indestructible.

GENERAL SPECIFICATIONS

Height of post from bottom of base to Signal 8 feet. Comes all wired and ready to mount or set. Can be mounted on small concrete foundations set in parkways or bolted to sidewalks or any concrete surface.

Simplex Interior Construction

Modernistic design with all aluminum parts integrally molded together. No other traffic signal even approximately equals SIMPLEX construction which is protected by U. S. Patent Applications, covering all the special features of design embodied in the housing, sealing, light partitions, lamp mountings, lens protection, etc.

Simplex Traffic Signal Lens

All Simplex Traffic Signals have special Simplex Lenses which bear on the inside surface a patented prism construction having the appearance of fish scales, but technically termed "Ungulae." Each prism directs the light downward and horizontal, thereby conserving the rays that would otherwise be wasted above the axis, and increasing the intensity and spread in the lower region where the light is most useful.

Between the "Ungulae" are flat spots which permit a portion of the light beam to pass straight through the glass, and thereby provide a long range, high intensity, attention-compelling beam along the street. This type is most useful under extreme and unfavorable sunlight conditions and where signals are spaced long distances apart. Simplex lenses bear no letters.

No. C257 Simplex Signal Traffic Signal

No. C262 Simplex Traffic Signal

No. C262. SIMPLEX 3-Colors, 1-Way Signal Complete with Top, Mounting Bracket, Steel Post, Cast Base, all wired ready for installation and connection to current. Height overall 10 feet 8¼ inches. Shipping weight about 136 lbs.
Each............................ **$46⁷⁵**

No. C264. SIMPLEX 3-Colors, 2-Way Signal Complete with Top, Mounting Bracket, Steel Post, Cast Base, all Wired ready for installation and connection to current. Height overall 10 feet 8½ inches. Shipping weight about 140 lbs.
Each............................ **$59⁴⁰**

No. C257. SIMPLEX 2-Color, 1-Way Signal Complete with Top, Mounting Bracket, Steel Post, Cast Base, all Wired ready for installation and connection to current. Height overall 9 feet 10 inches. Shipping weight about 128 lbs.
Each............................ **$39⁵⁰**

No. C257T. SIMPLEX 2-Color, 2 Way Signal Complete with Top, Mounting Bracket, Steel Post, Cast Base, all Wired ready for installation and connection to current. Height overall 9 feet 10 inches. Shipping weight about 134 lbs.
Each............................ **$49⁷⁵**

ALSO FURNISHED FOR OVERHEAD WIRING

If it is impractical or too expensive to tunnel and push conduit pipes under pavements and streets, we will furnish the above Signals with overhead wiring connections at the same prices. If that is your preference, specify overhead wiring when ordering.

☞ FOR OPERATING CONTROLS SEE PAGE 6

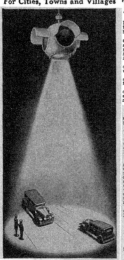

Flexible Rubber Safety Traffic Guides

Conforms with specifications of National Conference on Street and Highway Safety

A flexible rubber panel 22 inches long, standing upright in the pavement, 8½ inches high, with letters of Federal Yellow rubber vulcanized into the black rubber background. Cleated base of creosoted wood with metal anchor plates entirely enclosed in rustproof case.

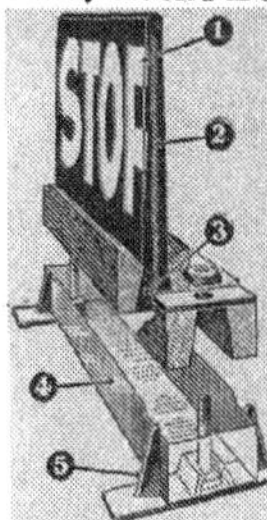

This rubber marker is embedded in the pavement in the middle of the street, directly in line of driver's vision. It may be easily read for a distance of 500 feet in the daytime and, being directly in the focus of the headlight, for a distance of 250 feet at night.

EASY TO INSTALL

Simply cut a slot in the pavement from 4 to 5 inches wide and 26 inches long; cut through the top layer of pavement without disturbing paving foundation. Insert the marker in this slot and pour concrete around it which, when set, will hold this signal in place.

BUILT FOR YEARS OF SERVICE

1. Thick letters of yellow rubber are inlaid—cannot wear off.
2. Fabric reinforcement insures upright position.
3. Wedge shaped base holds rubber slab firmly in place.
4. Non-rust case embedded permanently in pavement.
5. Metal anchor plate and bolt—refilling is easy.

EASY TO REPLACE

To replace marker at any time, unscrew nuts at each end, lift rubber marker and wooden clamps out and insert new marker, tighten nuts.

In ordering replacement markers send for "refills" only. It takes but a few minutes to replace these signs.

No. BG19. Flexible Rubber Guides. Worded STOP or SLOW on one side. Complete, each... **$6.75**

Rubber Refills for above, $5.85 each.

Complete signs worded SCHOOL SLOW, DANGER or ONE WAY on one side, each, $7.85. Refills for same, $6.95 each.

Complete signs worded NO-U-TURN, NO-LEFT-TURN, NO PARKING or SAFETY ZONE on both sides, each, $8.85. Refills for same, $7.95 each.

All other wordings such as HOSPITAL QUIET, STOP BOULEVARD, STOP R. R. XING, KEEP RIGHT, RIGHT TURN, are made special to order. Worded one side, $9.35 (refills, $8.45). Worded both sides, $10.35 (refills, $9.45).

NEW REFLECTOR BUTTONS

Weatherproof, indestructible, rhodium plated reflectors, with a metal cap. Easily installed in any signs.

Order No.		¾"	[illegible]	[illegible]	[illegible]	½"
	White or Amber	12c	11c	10c	10c	9c
E37D	Red or Green	14c	13c	12c	12c	11c

When ordering specify whether to be mounted in wood or in metal.

One Accident Prevented Pays for Many of Our New Safety Flags

There is no better way of giving warning to the motoring public of danger ahead than the Red Flag for day use or a Red Reflector at night. The Safety Flag will do the job day or night because it has the Red Flag and the Reflectors. It can be placed in the middle of the pavement, on the sidewalk, or any place where warning is required. Many towns use them at school crossings, where they are effective.

High enough to be seen and low enough not to be injurious. The strongest wind cannot blow it over. Always upright. Size of flag 10½ in. x 14½ in. Size of reflectors 1¼x1¼".

With self-righting base. No bulbs, no wiring, no oil. Always ready for use. Visible day or night 500 feet or more.

No. D808. Safety Warning Flags, each complete with 6 red 1¼" reflectors and base, any wording shown below, only.......... **$4.25**

DANGER
DANGER TRUCKS
DANGER WRECK
WET PAINT
KEEP OFF
DANGER DETOUR
GRADER AHEAD
DANGER
MEN WORKING
FRESH OIL
SCHOOL ZONE
DANGER AHEAD
CATTLE CROSSING

New Safety Warning Sign

Protects your men out on the job against the increasing hazards of heavier and faster traffic. Also a protection for the public.

Solid black 5½" letters on 20"x20" background of brilliant traffic yellow.

Light weight, yet extremely durable—the legs being made of ⅜" high carbon steel.

Each sign is equipped with two convenient handles for adjusting the angle of the legs. These handles, being hollow, also serve as flag sockets. Each handle is also equipped with a handy lantern lock.

Stands 43 in. high, width 28 in., wght. 23 lbs. Folds compactly into size 28x28x1¼ in.

No. D807. Men Working Sign, complete, only. **$5.95**

FOR MAKING YOUR OWN REFLECTOR SIGNS

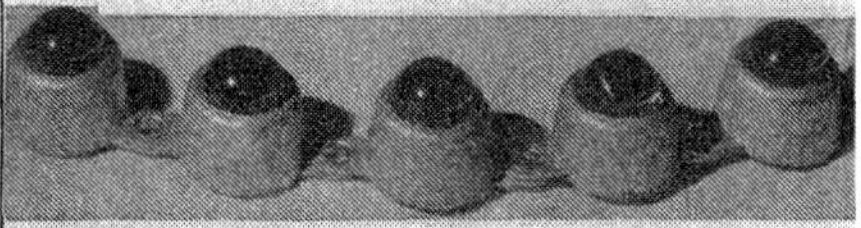

Cataphote reflector buttons in chains are easily installed in any signs to make them effective for night. Your local sign man can make the signs or lay out the lettering, and you can quickly attach the button chain. In the link hinges, there are screw holes for special theft proof screws. For either steel or wood signs. Chains Cadmium plated. We recommend chains painted same color as lettering. Unless otherwise specified, all buttons shipped separate from chains to permit dipping or spraying.

ONLY $1.35 Per Foot

Modernize present daytime signs and double their effectiveness. Reflector button chains will make your daytime signs effective at night, when most needed, because 6 times more accidents occur at night than during the day.

Cataphote Buttons are optically perfect and permanently brilliant. Angularity of lens 100 degrees. ☛ Your choice of colors, crystal, yellow, red or green.

The No. D805 Reflector Chain has ½ in. Cataphote reflector buttons, center to center spacing 1⅛ in., with 11 buttons per foot. The No. D806 Reflector Chain has ⅝ in. Cataphote reflector buttons, center to center spacing 1⅛ in. or 2 in. with about 9 buttons per foot.

When ordering ☛ specify Cat. No., color, how many feet you want, and whether for steel or wood signs. ☛ For No. D806 also specify spacing: 1⅛" or 2".

No. D805 Reflector Chains, ½" buttons, per foot.......................... **$1.35**

No. D806 Reflector Chains, ⅝" buttons, per foot.......................... **$1.65**

☛ **NET PRICES**
Less 3% Discount for Cash or C. O. D. Orders

TRAILER FLAG only $1.50

Size of Flag 10½" by 14½"

With bridle snap for fastening to back of truck on load and with timber drift for driving into end of pole or logs.

No. D809. New Trailer Flag, red with white lettering, each **$1.50**

100 HOUR

Utility Lantern

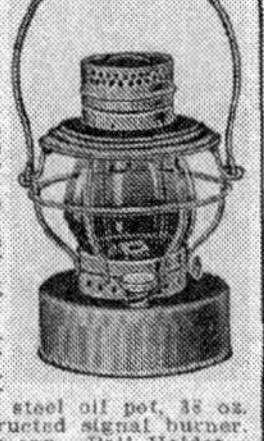

The very finest utility lantern for municipal street use. Ruggedly made. Approved by Interstate Commerce Commission for railroad service which means it's built to strict specifications other lanterns can't meet.

Will burn for over 100 hours without attention. Railroad signal lantern ventilation. Will stay lighted in any weather. Genuine ruby globes or any color desired.

Built for rough service. Will outlast other lanterns by several years. Steel frame, electric welded wire guards. Drawn steel oil pot, 38 oz. capacity. Heavily constructed signal burner. Large filler with convex cap. Bail Holder—keeps bail in vertical position when desired. The most economical and most dependable warning signal made.

No. D810 One Hundred Hour Utility Lantern, each.......................... **$1.79**

No. D811 One Hundred Hour Utility Lantern, per dozen.................. **$17.90**

$6.55

RED Reflecting Signals

YOUR CHOICE RED AMBER WHITE

Price does not include posts

Our 30 button CATAPHOTE Reflecting Signal is something distinctively new and is the latest development in Reflecting Signals. It is, without doubt, most EFFICIENT, most DURABLE.

In the 8½-inch circle are inserted 30 HIGH POWER ⅞-inch reflecting buttons, hermetically sealed in the Signal, making them rain-proof, dust-proof and weather-proof.

Easily placed because they can be screwed, or bolted to inexpensive 3x3 wood posts, or to steel pipe and angle iron posts, or to any flat surface. Mounted on 9x9 inch back plate.

No. C815. Reflecting Signals, each **$6.55**

No. D902. Same with 18x18 in. back plate lettered STREET END—DEAD END or any other short warning notice you desire, black letters on yellow background, each only $8.55

Flexible Rubber Traffic Guides

THE NEW KIND 🖝 EASY TO INSTALL

A flexible rubber panel 18 inches long standing upright on the pavement 7 inches high, with bold letters of Federal Yellow moulded in the sign.

Easily read for a distance of 500 feet in the day time and being set directly in the focus of headlights; for a distance of 250 feet at night.

🖝 Made with STEEL BASE to BOLT to pavements. No slots to cut and installed without defacing the roadway or street.

No. C212. Flexible Rubber Guides. Complete, each.......... **$6.50**

🖝 Your choice of any design shown.
No. C213. Refill Markers, each........... $4.00

Size of sign 7x14 inches.
Size of base 4½ inches by 18 inches.
Packed 2 signs and bases each box.
Net shipping weight per box 25 pounds.

SLOW   **NO U TURN**   **SCHOOL SLOW**

They will never tax taxpayers' dollars. They cost little and nothing to maintain. Always in line with the driver's eyes.

These Silent Policemen arrest every driver's attention. They get willing obedience.

Every community can afford them. They require no mechanical or electrical connections. Simply fasten the anchor bolt in a little cement and you have a permanent installation.

There is nothing about them to wear out or go wrong. High enough always to form an effectual safeguard, low enough not to damage, or be damaged by a car hitting them.

Hundreds of towns are using these stop plates for their stop streets; are placed on ground in center of street; are 22 inches long, 14½ inches wide and 3½ inches high at base. Worded STOP or SLOW SCHOOL, NO LEFT TURN, KEEP TO RIGHT, BOULEVARD STOP, RAILROAD STOP, ARTERIAL STOP, SLOW DANGER, SLOW DANGEROUS CROSSING.

No. C820. Plates, each........... **$5.62**

Alloy-Aluminum Sidewalk Auto Signs

🖝 The Aristocrat of its Class 🖜

🖝 Size 10x13 Inches. Stand 36 Inches High

These are made of our genuine hard alloy-aluminum metal. 🖝 Guaranteed never to TARNISH. 🖝 Not to RUST. 🖝 Never to BREAK. 🖝 Not to CORRODE. They are indeed the brightest and finest Parking Signs in the world.

Other signs there are HIGHER IN PRICE, but being comparable to these in quality. Our sign factory in Chicago where Alloy-Aluminum signs are made exclusively is, we believe, not only the largest, but the best equipped in the country. 🖝 We aim to quote the LOWEST PRICES to our customers who order direct from this catalog.

No. 774. Alloy-Aluminum Parking Signs complete with 16-lb. cast iron base, 10-inch diameter, with inch steel pipe to stand 36 inches high. Each....... **$6.95**

🖝 Many customers want the Sign Plates only and supply their own 1¼-inch pipe cut just the length they want it and make holes in sidewalks so the pipe can be cemented in. Such signs can not be carried away or moved. If wanted set 7 feet above pavement, it can easily be done at small expense. We quote below on Sign Plates only.

No. 677. Alloy-Aluminum Parking Signs. Plates only. No pipe and no base. Each........... **$4.95**

🖝 Remember to give the wording you want. 🖜

HALF HOUR PARKING

NO PARKING FIRE HYDRANT
Lettered Both Sides

NO PARKING POLICE ORDER

🖝 Notice. For 50c extra per sign we will make the No. 774 with the SPEAR POINT BASE.

SPEAR Point Base
SPECIAL — See That 🖝

Heavy Cast Iron Spear Base, 21 inches long. Weight, 22 lbs. Drive in ground and it holds Sign like an anchor. Can't be pulled out.

No. B303. Any Alloy Aluminum sign with this Spear point base. Each.... **$6.95**

🖝 Spear points only, no sign included..... **$2.05**

PARK HERE

(Keep to Right)   (Stop)
(Reserved for Fire Dept.)
(No Left Turn)   (One Way Street)
(Keep 20 Feet Away)
(Street Closed)
(School—Slow)
(Reserved for Police)
(Detour 🖝)
(Detour →)
(No Parking Between Signs)

ONE WAY STREET

No Extra Charge For Any Wording Desired

3% Off CASH OR C. O. D.

REFLECTOSTRIP

Weatherproof and theftproof. Reflectors cannot be unscrewed or unbolted from the strip itself. Most compact reflective unit made, only ½" thick. Can be bent to a radius of 4".
No. E837. In 5 foot lengths, per foot............ **$1.40**
Less 15% for 20 feet or more.

With these handy reflectors in strip form, you can equip every danger spot with reflective warnings visible for over 120 feet. No special tools are necessary to put these strips in place ... simply a hack saw and screw driver. Drilled and scored every foot. Easily installed anywhere.

REFLECTOSTRIP consists of highly efficient reflecting units, securely fastened to strips of 18 gauge Parkerized Rust-Resisting Steel, 1⅝" in width, eight reflectors to a foot. Strips can easily be cut into any length desired. Each five foot section comes complete with 20 Parkerized Rust-Resisting Screws. Red, Green, Crystal and Amber are standard colors and Orange, Yellow, Blue and Aquamarine are supplied on special order. Each reflecting unit is made up of seven clustered glass lenses and the entire unit has an optical diameter of ⅞". Units are set 1⅝" from center to center and at a distance appear as a continuous line of light. Will give a clear reflection up to 80 degrees.

11

United States Standard
HIGHWAY AND STREET SIGNS *for* $1 69 EACH

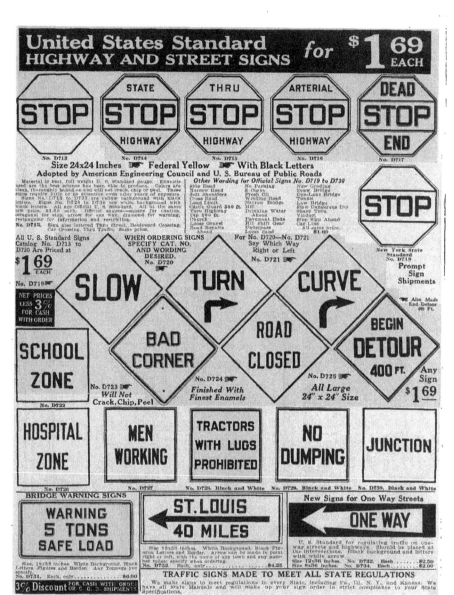

STOP — No. D713

STATE STOP HIGHWAY — No. D714

THRU STOP HIGHWAY — No. D715

ARTERIAL STOP HIGHWAY — No. D716

DEAD STOP END — No. D717

Size 24x24 Inches ☞ Federal Yellow ☞ With Black Letters
Adopted by American Engineering Council and U. S. Bureau of Public Roads

Material is best, full weight U. S. standard gauge. Enamels used are the best science has been able to produce. Colors are clean, thoroughly baked on and will not crack, chip or peel. These signs require little or no attention even after years of exposure. Signs No. D713 to D737 are yellow background with black letters. Signs No. D728 to D730 are white background with black letters. All are Official U. S. standard. All at the same low price, $1.69 each. Official shapes—square for caution, octagonal for stop, arrow for one way, diamond for warning, rectangular for information and restriction.

No. D713. Sign also lettered Thru Street, Boulevard Crossing, Car Crossing, Thru Traffic. Same price.

All U. S. Standard Signs Catalog No. D713 to D730 Are Priced at
$1 69 EACH
No. D719 ☞

Other Wording for Official Signs No. D719 to D730

Side Road	No Parking
Narrow Road	S Curve
Soft Shoulders	Fresh Oil
Cross Road	Winding Road
Load Limit	Narrow Bridge
Castle Guard 300 ft.	Hill
Turn Highway	Drinking Water
Dip 300 ft.	Ahead
Church	Pavement Ends
Loose Gravel	Hill Shift Gear
Road Repairs	Underpass
Ahead	Loose Sand

New Grading
Draw Bridge
One-Lane Bridge
Tunnel
Low Bridge
Slow Dangerous Dip
Sharp Turn
Viaduct
Stop Sign Ahead
Car Line
All same price. $1.00

STOP — New York State Standard No. D718

Prompt Sign Shipments

WHEN ORDERING SIGNS SPECIFY CAT. NO. AND WORDING DESIRED. No. D720

For No. D720—No. D721 Say Which Way Right or Left No. D721 ☞

NET PRICES LESS 3% FOR CASH WITH ORDER

SLOW

TURN

CURVE

☞ Also Made End Detour 400 Ft.

SCHOOL ZONE — No. D722

BAD CORNER — No. D723 ☞ Will Not Crack, Chip, Peel

ROAD CLOSED — No. D724 ☞ Finished With Finest Enamels

BEGIN DETOUR 400 FT. — No. D725 ☞ All Large 24" x 24" Size

Any Sign $1 69

HOSPITAL ZONE — No. D726

MEN WORKING — No. D727

TRACTORS WITH LUGS PROHIBITED — No. D728. Black and White

NO DUMPING — No. D729. Black and White

JUNCTION — No. D730. Black and White

BRIDGE WARNING SIGNS

WARNING 5 TONS SAFE LOAD

Size, 18x36 Inches. White Background, Black Letters Figures and Border. Any Tonnage you specify. No. D731. Each, only............$6.00

☜ ST. LOUIS 40 MILES ➜

Size 15x56 Inches. White Background, Black Figures Letters and Border. Arrow can be made to point right or left, with the name of any town and any number miles; specify when ordering. No. D732. Each, only............$4.25

New Signs for One Way Streets

☜ ONE WAY

U. S. Standard for regulating traffic on one-way streets and highways. Should be placed at the intersections. Black background and letters with white arrow. Size 12x36 Inches. No. D733. Each.......$2.50 Size 9x36 Inches. No. D734. Each.......$2.00

TRAFFIC SIGNS MADE TO MEET ALL STATE REGULATIONS

We make signs to meet regulations in every State, including Pa., Ill., N. Y., and Kansas. We have all State Manuals and will make up your sign order in strict compliance to your State Specifications.

3% Discount FOR CASH WITH ORDER OR C. O. D. SHIPMENTS

12

NEW Street Traffic Line Markers
BUILT FOR HEAVY DUTY SERVICE
LAYS STRAIGHT TRUE LINES WITH ANY PAINT

The most practical type of marker made. Paints a side stripe (operator walks naturally and does not have to "straddle" newly painted line). Especially designed for painting safety lines on streets, highways, parking spaces, in warehouse and industrial plants, etc. This is a one man machine. Will paint over a mile of lines an hour. No "trick gadgets" to get out of order.

Simple gravity feed. Your paint flows in an even, smooth line from reservoir to the all-metal Flexible Spreader which lays down a fine, clean cut line on any type of road no matter how bumpy the surface may be. The operator has complete control of the line density from a control handle on the push bar. Even when full of paint our new TRAFFIC LINE MARKER with its roller bearings and rubber tired wheels is light and easy to handle. All metal, flexible spreaders do the actual painting. No brushes, wheel bands or spraying apparatus to get out of order. Spreader is detachable for cleaning and also to attach a different size spreader as desired. The flexible spreader is always ready for service if placed in just sufficient kerosene oil to cover it when not in use.

Tank capacity: three gallons. Comes complete with Spreader of any width (when ordering please specify 2, 3, 4, 6 or 8 inch Spreader). Will mark up to within 1½ inch of any parallel obstruction. Rubber tired wheels with roller bearings. Automatic road depression leveler provides for dips and bumps and the wheels roll the paint right into the road.

For efficiency, dependability and economy, our new Traffic Marker has no equal, and the labor, time and money it saves justifies its purchase.

No. B678, New Traffic Line Marker......... **$50⁰⁰**

☞ $48.50 cash with order or C. O. D.
Specify size of spreader wanted—2, 3, 4, or 8 in. width.

$50
SAVES ITS COST
This Marker soon saves all it costs in labor alone.
One gallon of good paint will mark approximately 800 linear feet of line 6 inches wide.
Uses any zone paint or lacquer.
Try it 30 days at our risk.

30 DAYS FREE TRIAL
Guarantees Your Satisfaction
Any Official is welcome to order the New Traffic Line Marker on 30 Days' Free Trial. If it fails to satisfy you after testing it thoroughly on your own work, return it. No ifs, ands or buts to this offer.
☞ And we will pay the transportation charges both ways in event of dissatisfaction.

Special Municipal "Wearproof" *QUICK DRYING* Zone Marking Paint $1 50 PER GALLON

Wearproof Zone Paints Are Ready Mixed Paints
Suitable for any marking machine or for hand brush use

"Wearproof" is guaranteed paint. Extra high grade when compared with ordinary paint. Not only covers more surface and spreads farther, but dries quicker with a higher luster and the pigments possess greater brilliance. In formulating this paint our chemist has produced a paint that will insure extra long life so that when the time comes for re-painting a good foundation remains for the next line.

MAXIMUM COVERAGE AND DURABILITY
Actual tests by municipalities have proved that our TRAFFIC Paint gives more coverage per gallon than other paints. It will cover a strip 450 to 750 feet long and 6 inches wide, with each gallon, but coverage must be figured after taking into consideration type of marking machine, kind of road, and weather conditions. We claim the maximum of durability, and invite actual traffic tests against any other paint. No "pick up" or unsightly smears. Our TRAFFIC Paint will last longer and go farther.
☞ Especially penetrating on concrete, brick, cement and asphalt streets. The liberal spreading capacity of "Wearproof" makes it the most economical. Easy to apply and flows out smoothly.
Used and re-ordered by hundreds of Towns, Cities and Villages on streets and highways for marking safety zones, traffic lanes, parking spaces and pedestrian crossings. Painted lines guide traffic, help school children and others to cross in safety and the more you use, the fewer accidents will happen.
"Wearproof" Zone Paint retards the absorption of stains, oils and greases and prevents the disintegration of concrete by abrasive wear, walking and trucking. ☞ We ship our customers direct from our paint factory in Chicago, and as jobbers, dealers and salesmen are eliminated, our customers save about 40% on all paint orders.
"Wearproof" Zoning Paint can be wiped off and cleaned like tile, and will not crack, chip or scale, and is not affected by acids, alkalies, gases or the caustic elements of brick or asphalt. Comes ready for use without using any thinner.

WHITE - RED AND YELLOW

LOW PRICES IN CANS OR BARRELS

Low Prices ☞ In Barrel Lots
White	$1.50 per gallon
Yellow	1.80 per gallon
Red	2.10 per gallon

☞ Special 30-Gallon Barrel or the regular 50-Gallon Barrel, same price. DISCOUNT 3% For Cash with Order or C. O. D. Shipment.

Low Prices ☞ In 5-Gallon Cans
White	$ 8.65 for full can
Yellow	10.15 for full can
Red	11.65 for full can

Low Prices ☞ In 1-Gallon Cans
White	$1.88 for full can
Yellow	2.20 for full can
Red	2.50 for full can

Telegraph W. S. Darley & Co. any time about anything.

15

Municipal Tar and Asphalt Melting Kettles

for Municipal, County and State Highway Maintenance

$155

50 Gal.
Capacity
Complete

Simple to operate. After filling the fuel tank, a few minutes' pumping will produce from 40 to 50 lbs. pressure, sufficient to operate burner and kettle continuously for several hours. Air is used only to force fuel to the burner. When pressure gets low, a few strokes of the pump raises pressure for several additional hours' operation. The burner is preheated for five minutes (instructions are furnished with each kettle) and equipment is ready for operation.

All-Steel Construction in every way.
SAVES FUEL because kerosene or coal oil is cheaper than wood.
SAVES TIME because kettle can be put in operation in 5 minutes.
SAVES LABOR because kettle requires very little attention after it is started.
NO SMOKE—City engineers recommend these kettles because the smoke nuisance, sparks, etc. are entirely eliminated.

Fuel tank of seamless pressed steel, and brazed, tested at 300 lbs. hydrostatic pressure. Tinned inside and outside to make it rustproof. Pressure gauge, fittings and valves of best materials and workmanship available.

Strapped in position, therefore ready and quickly removed and used, with burner and hose, as portable equipment for drying, thawing, general heating and repair work.

Number	Capacity	Oil Tank	Fuel per Hour	Ship. Wgt.	Price
No. D908	50 gals.	20 Gal.	2 Gal.	440 lbs.	$155.00

Steel wheels standard. Hard rubber tires $35.00 extra. Pneumatic rubber tires $45.00 extra.

Patrol Patching Heaters and Melters

For heating asphalt, pitch, tar, for expansion joint filling and small patch work. Hot tar or asphalt in 10 minutes. Burns kerosene or furnace oil. Heats twice as fast as a wood fire—no waiting, absolute temperature control—no sales, no sparks.

The No. D647 Patrol Heater has steel melting tank, dished bottom, with 1½-inch draw-off cock. The No. D648-D650 Melting Furnaces are same construction but do not have draw-off cocks and are made with pots of cast iron which don't coke material. For melting leadite and other compounds used by Sewer and Water Depts.

Heavy sheet steel furnace, 18-inch roller bearing steel wheels—2-inch face. 5-gal. welded steel oil tanks. Quick acting brass pump, pressure gauge, fittings and valves are of best materials and workmanship. Special oil resisting rubber hose with ground brass unions. Easy to start burner. After filling tank a few minutes pumping will produce 30-40 lbs. pressure, enough to operate burner for several hours. Air used only to force fuel to burner.

Heat easily controlled by simply sliding burner on rail forward for more heat and pulling it back for less heat. Gives an intense, reddish blue, clean and steady flame 1800° to 2000° F. No smoke.

No. D647. Patrol Patching Heater, capacity 10 gal., shipping weight 190 lbs. . . . **$80.00**

No. D648. Compound Melting Furnace, capacity 5 gal., shipping wt. 225 lbs. . . . **$80.00**

No. D649. Compound Melting Furnace, capacity 15 gal., shipping wt. 315 lbs. . . . **$90.00**

No. D650. Compound Melting Furnace, capacity 25 gal., shipping wt. 445 lbs. . . . **$100.00**

Pouring Pots and Joint Fillers

All our pouring pots and joint fillers are light and strong, with all welded construction. Extra durable. Accumulation of cold asphaltic material may be burned out, as no solder, lock-seaming or galvanizing are used in their assembly. All seams acetylene welded.

No. D830 and D831 Crack and Joint Fillers have three size of pouring pots and a quick setting ground seat shut off valve designed to hold the heat and avoid congealing. No packings are used in the valve assembly and these pots may therefore be burned out without damage. No. D831 will fill under low heating kettle draw offs. Filler screens available when specified. Neat and accurate pourers, they do better work and frequently save their cost in material in a single day. Wght., 7 lbs. capacity, 2 gals.

No. D830. Joint Filler, each . . . **$10.00**

No. D831. Joint Filler, each . . . **$11.00**

D830   D831

Patented adjustable opening spout for light or heavy pouring.

For sheet pouring hot asphalt and tar in road construction, waterproofing and repairing. Wgt., 6 lbs. Capacity 3 gals. Spout, 3 in.
No. D829. Each . . . **$6.25**

For pouring expansion joints and cracks in pavement. Patented nozzle with a size openings and quick removable basket filler screen. Wght., 6 lbs. Capacity, 4 gals.
No. D828. Each . . . **$6.25**

Combination Melting Furnace and Detachable Torch

$99⁷⁵

Melts
250 Lbs.
Lead
In 18
Minutes

Both
Furnaces
Are
Complete
With
Portable
Torch
Shown
on
Opposite
Page

Complete As Shown
Burns Kerosene or Coal Oil

For melting lead and soft metals. Torch instantly detachable for preheating before welding and expanding for sewer and gas work, for melting lead in furnace, heating frozen water pipes, frozen ground, melting ice and snow, for wood burning in summer. For heating salamanders, asphalt and tar road kettles, roofers' kettles with small pour and low fire boxes, water, asphalt, compounds and soft metals. Melts material in all the uses required by wood.

The detachable torches supplied on these Melting Furnaces are the Portable Heating and Melting Torches, shown and described above. This combination outfit at our low price is a super value, in reality two outfits for the price of one.

Furnace made of heavy sheet steel, ⅛ in. thick. Furnished with pot rack and bar. Complete with cast iron melting pot.

The Melting Furnace on wheels has 24 in. roller bearing steel wheels with 3¼ in. face. The No. D839 Outfit will melt 490 lbs. of lead in 25 minutes and keep it molten at a few cents per hour.

Catalog No.	D838	D839	D840
Capacity of melting pot, lbs.	250	450	750
Time required to melt full pot of lead, minutes	18	25	35
Fuel consumption, gals. per hr.	1¼	2	2
Flame dimensions, inches	3x30	3½x30	3½x30
Fuel tank capacity, gals	10	10	15
Price, ON LEGS	$99.75	$116.25	$131.85
Shipping weight, lbs	300	335	375
Price, ON WHEELS	$125.00	$140.00	$165.00
Shipping weight, lbs.	350	385	425

A CHILD'S LIFE MAY BE SAVED BY THIS TRAFFIC SENTINEL

$9⁸⁵

Easily movable to the center of the street and taken away when not in use. The school janitor or an older boy can take care of it.

Our streets today are filled with too many stop signs. The Sentinel, however, is placed in front of the motorist only when he is supposed to heed it; namely, when the children are arriving at school in the morning, at noon and at closing time. It explains in bold letters the REASON why it is there. The motorist respects this reason and obeys.

Distinctive and snappy looking Sentinels command respect. Of substantial construction, they last for years and retain their good appearance. Solid and rigid, they remain where placed. Made of 18 gauge prime steel and finished in baked enamel. Color scheme: red uniform, yellow sign with black letters, white belt, black puttees. Will withstand severe weather. Figure is detachable from the heavy cast iron base. Height overall 5 ft. 3 in.

No. D539. Traffic Sentinel complete. . . **$9.85**
Same as above but double faced, two overall 5 ft. 3 in. . . . **$16.25**

Figures back to back.

Your choice of lettering: SCHOOL DRIVE SLOWLY or PLAYGROUND DRIVE SLOWLY or ZONE OF QUIET. Special lettering $1.50 extra per sign unless order is for 12 signs, same wording.

SCHOOL
DRIVE
SLOWLY

FOR MARKING PARKED CARS

Our tire marker paint pencil is the most efficient means of keeping track of parked cars to get the overtime parkers. Marks tires plainly. Rainproof. Red, white, blue, yellow, green. 7 inches long.
No. E546. Per dozen only . . . **$1.25**
In lots of 6 doz., 95c per doz.
No. E547. Metal holders, pocket style, each . . . 15c
No. E548. Wood holder, long extension style, each . . . 90c

16

$8.95

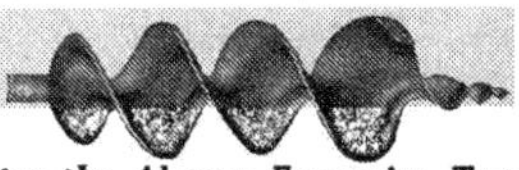

NEW AUGER FOR BORING UNDERGROUND TO LAY PIPE

Eliminates the Always Expensive Tearing Up of Streets, Pavements, Railroad Intersections

Our new Auger will bore underground through clay, earth, sand and gravel for laying pipes under streets, lawns, highways, sidewalks, embankments, railroad crossings, etc. Bores perfectly clear holes rapidly and at a very low cost. Saves the great expense and delay of tearing up paving and trench digging. A wonderful time and money saving tool for every municipal dept. for laying services from house to main as shown at right. ☞ Made for boring underground to lay all sizes of pipe from 1½ in. to 6 in.

Enormous saving can be made with an Auger in so many different ways. For example, it can be used for draining ponds, draining dykes, making connections under rivers and creeks, draining cellars, detecting leaks, making soundings, laying pipe under hedges, parks, gardens, buildings, reservoirs, race tracks, constantly occupied storage yards, and so forth. With an Auger you can put in more pipe in 15 or 20 minutes than a man can dig ditching in an hour.

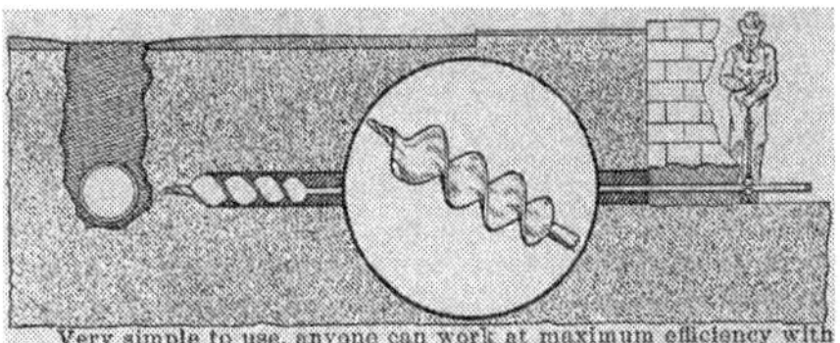

Very simple to use, anyone can work at maximum efficiency with the new auger. Any unskilled worker can use it. Light in weight, portable, easy to carry and handle. As shown above, the operator merely turns the auger to the right with a pipe wrench until the twisted part is filled with dirt, then he pulls the tool out and brings out a load of earth with it. The auger twists are cleaned and the tool is ready to go back in the bore for another load.

A special auger designed and made expressly for this service. Made of an alloy of steel, the product of the most advanced scientific and metallurgical engineering. HEAT TREATED. The small diameter cutting head has a small double twist which serves as a pilot and also enables the auger to gradually bite into the ground. The large diameter of the auger is sized according to pipe to be laid and is made with large open double twist to allow for dirt to be removed. Gradually tapered down to avoid unnecessary friction and to make boring easier and quicker. The shank of the new auger is threaded for standard pipe so that by adding sections of pipe, boring can be done with amazing exactness 50 to 60 feet.

Showing how the new Auger saves expense of destroying paving, cost of new material, labor, traffic delays.

The above picture is full of meaning for every Supt. It is expensive and difficult to remove pavement and replace it, and it is conservative to say that one job like this will save several times the cost of the new Auger.

Catalog No.	E751	E752	E753	E754	E755	E756	E757	E758	E759	E760	E761	E762	E763	E764	E765
Pipe Size	1"	1¼"	1½"	1¾"	2"	2¼"	2½"	2¾"	3"	3¼"	3½"	3¾"	4"	4½"	6"
Price, each	$7.70	$8.30	$8.95	$9.60	$12.60	$13.45	$13.80	$14.70	$15.75	$17.95	$19.90	$22.80	$29.97	$35.80	$67.40

THE MOST EFFICIENT TEST PLUG IN THE WORLD

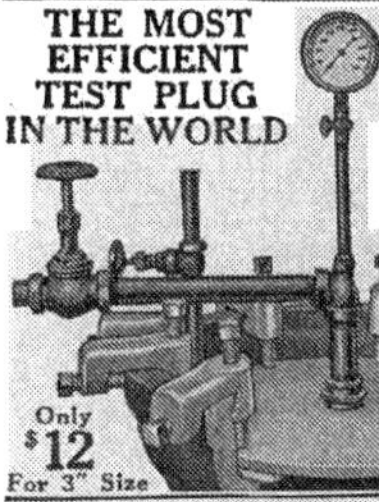

Only $12 For 3" Size

For Testing All Mains Easily and Quickly

Used by City of Chicago

Easily installed by unskilled laborers. A steel plate, a rubber gasket, together with the required number of clamps, all quickly installed by one man, makes the pipe line ready for any test. Guaranteed for a hydrostatic pressure of 300 lbs. per sq. in. Plates are heavy pressed steel, clamps are alloy steel with case hardened set screws.

The plug is also very useful where a pipe line has broken and it is necessary to restore service on the main temporarily.

Each test plug is tapped for ¼-inch pipe connection to release air from the main. Also tapped ¾ inch to 1½ inch for Test Pump connection, and for flushing of trenches. Gauges not included.

Designed to Save You Money

The clamps may be used for different sizes of pipe. Only a separate steel plate and gasket are required for each pipe size. For example, if you purchase a Plug for testing 4-in. bell end pipe and later desire to test 6-in. bell end pipe you have only to purchase the steel plate and gasket, 6-in. size, and one more clamp. ☞ Write for prices on plates, gaskets and clamps.

FOR BELL END OF PIPE

Catalog No.	D903	D904	D905	D906	D907	D908	D909	D910
For pipe size, inches,	3	4	6	8	10	12	14	16
Price, complete	$12.00	$15.00	$22.00	$28.00	$35.00	$42.00	$55.00	$65.00

FOR SPIGOT END OF PIPE

Catalog No.	D969	D970	D971	D972	D973	D974	D975	D976
For pipe size, inches,	3	4	6	8	10	12	14	16
Price, complete	$20.00	$29.00	$30.50	$40.00	$50.00	$60.00	$72.00	$84.00

WRITE FOR PRICES OF LARGER SIZE TEST PLUGS

PORTABLE HEATING AND MELTING TORCHES

An All-Around Torch for Melting Lead, Thawing, Burning Weeds

$44 Complete

These are portable one-man outfits and burn kerosene (coal oil). The burner is a special pre-heating type, equipped with heating pan and can be started within three minutes. The flame is intensely strong and easily regulated as desired. Scale treated, improved, seamless coil generator produces flame 2000° F. Reddish blue flame burns clean and smokeless, no sparks, no ashes.

No compressor required, only necessary to pump up 40 to 50 lbs. of air in tank, and that is easy with the powerful hand pump, and it is ready for anything you have to do. Operates several hours with one pumping.

Most successful outfit ever offered to Waterworks men for melting lead out of joints and to fit joints.

This burner outfit will run the lead of any size joint in just a few minutes. Will do the work of five men thawing out pavement and frozen ground in winter; with the No. D834 Burner 6 cubic feet of frozen ground can be dug in 20 minutes. And there are any number of other uses, such as thawing frozen hydrants, services, mains, gate valves, melting snow, removing ice from man-hole covers, thawing coal car hoppers, switches.

☞ For heating salamanders, asphalt and tar road kettles, roofers' kettles with small, long and low fire boxes, water, asphalt, compounds and soft metals. Melts material in half the time required for wood. For preheating before welding, and expanding for shrink fits. ☞ For weed burning in summer.

The new portable Heater and Melter torch has pressed steel welded fuel tank. Outfit comes complete with powerful quick-action hand pump, oil-resisting hose (No. D832 and D833 Outfits—6 ft. length; No. D834 and D835—10 ft. length; No. D836 and D837—12 ft. length) with brass unions, 2 to 4 ft. extension pipe, and burner with combination oil needle valve and fuel strainer. Burner not affected by wind or cold weather. Burner fitted with adjustable shield. Tanks have side handles or bail.

Torch Outfits Complete With Tank, Pump, Hose and Burner

Catalog No.	D833	D834	D835	D836	D837
For what capacity melting pot, lbs.	250	450	750		
Time required to melt full pot of lead, minutes.	18	25	35		
For what capacity tar kettle, gals.	25	50	75	110	165
Fuel consumption, gals. per hour.	1½	1½	2	3	3
Flame dimensions, inches	3x30	3½x30	4½x30	4½x36	4½x42
Fuel tank capacity, gals.	5	10	15	15	20
Shipping weight, lbs.	48	80	110	115	125
Price, complete, each.	$44 00	$52 00	$62 50	$65.00	$72.50

TELEGRAPH ANYTIME!

Your message comes direct over our Customers Private Wire. For quick service any time telegraph us from your local office. Costs you nothing. Try it. ☞ Wire Charges collect of us.

Simplex Pipe Pushers Are Saving Time and Money
For Water, Gas, Electric Departments
All Over the World

Only 97^{00}
And 129^{25}

From start to finish of a pipe pushing job the multiplied man-power of Simplex Double Leverage Pipe Pushing Jacks saves time and money at every step. Trenching is reduced to a minimum and high cost of repairing and replacing broken streets, sidewalks, parkways, is eliminated. Initial cost of Simplex Pipe Pusher is very reasonable and is the only cost. Frequently, on one or two installations the savings will pay for the jack.

The cost of Simplex Pipe Pusher is very low when its economy and practicability are compared with the expense of destroying pavement, together with cost of material, labor, traffic delays.

Then, too, the Simplex will serve for hundreds of additional jobs, irrespective of weather or soil conditions. Each pusher is built to a high standard of mechanical excellence and is guaranteed for dependable, economical service.

Showing how Simplex Pipe Pusher saves expense of destroying paving, cost of material, labor, traffic delays.

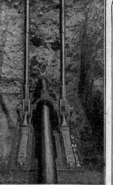

The Standard (capacity ¾ in. to 2 in.) and the Super (capacity, 2 in. to 4 in.) Simplex Jacks are designed for pushing pipe through the most unfavorable soil conditions without crushing or distorting the pipe. The practical improvements in these Simplex Jacks are based upon a research of the failures that heretofore have surrounded the use of pipe pushing jacks.

DOUBLE LEVERAGE

The Simplex Pipe Pushing Jacks can be operated by 2 or 4 men, depending upon soil conditions, and an exclusive and valuable feature is that when solid cribbing or blocking is difficult to obtain, it is then possible to hold the jack against the back pressure with one lever, while the other lever is being operated.

In the picture at upper right, the 3 tapered jaws are shown in position in the Simplex, and also after removal, alongside of the Jack.

The design of these tapered jaws insures the complete surrounding or gripping of the pipe whereby kinks or flattening of the pipe is overcome.

The two levers or sockets of the Simplex Jacks can be operated singly, alternately or together, depending upon the size of pipe, soil conditions and cribbing.

In overcoming the common tendency to flatten pipe, the extraordinary power of the Simplex Pipe Pushing Jacks can be safely utilized.

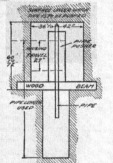

Illustrating a plan view of the Simplex Pipe Pushing Jack with a piece of pipe ready to be pushed through the earth, and a pilot point screwed on the end of the pipe.

This view also shows the position of the cribbing or wooden beam which provides a firm bearing for the base of the Jack; and the approximate size hole which should be dug at a depth of about 5 inches below the level at which the pipe is to be pushed.

Each size of pipe requires a set of tapered jaws, the reason being that perfect or true gripping is necessary to avoid crushing the pipe and in order that the great pushing power of the jack can be utilized.

Simplex Pipe Pusher has been especially designed for installing lead or iron pipe without digging, tunnelling or trenching. Of course, it is a known fact that lead pipe cannot be pushed because it will crush, but by using a piece of steel or iron pipe, the lead pipe can be drawn through and the result is a simple, speedy efficient installation with less effort and less expense. Only one excavation is necessary—namely, at the street main. Simplex Pipe Pusher is placed in basement in which an excavation is all ready and a piece of iron pipe is pushed out to the main as shown above. After iron pipe has been pushed to the main the lead pipe is attached to it, Pipe Pusher is turned around and iron pipe pulled out. Lead pipe follows in hole made by iron pipe, as shown below.

If soil conditions are of loose sandy type, another method of installing lead pipe is to push a piece of oversize iron pipe to the main and slide the lead pipe through by hand, anchor the lead pipe and pull out the iron pipe, leaving the lead pipe in.

No. B 83 Standard Simplex Pusher for ¾", 1", 1¼", 1½", and 2" pipe. ☞ With 2 Levers and 2 Extension Levers and ☞ 1 set of any size Pipe Jaws. Weight of Jack 146 lbs., without accessories. **$97.00**
☞ Extra sets of jaws for the above, per set............$9.60

No. B 84 Super Simplex Pusher, for 2", 2¼", 3" and 4" pipe. ☞ With 2 Levers and 2 Extension Levers and ☞ 1 set of any size Pipe Jaws. Weight of Jack 211 lbs., without accessories. **$129.25**
☞ Extra sets of jaws for the above, per set............$12.00
(OUR PRICES include a Special STEEL PILOT with every set of jaws for 2" and larger.)

NEW Champion Hydraulic Pipe Pushers

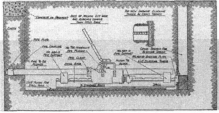

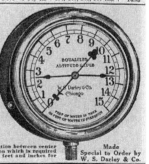

Long Distance Electric Alarm Gauge Panels

For Water Tanks, Stand Pipes, Towers or Reservoirs

The Long Distance Electric Alarm will notify you when stand pipe, tank or reservoir water supply is at a low level and pumping should be started, also when the water reaches a high level and pumping should be stopped. It is a HIGH-LOW WATER ALARM. A very reliable system —accurate, practically indestructible, very much less expensive than other apparatus designed for the same service,—not only in initial cost but installation and maintenance as well.

Regardless of distance between stand pipe (gauge location) and pump house (panel board location) the installation can be completed easily and inexpensively. Only one overhead wire is required between these two points, the circuit being completed by using pipe line as ground connection in place of a second wire. The aerial can be ordinary telephone wire. Telephone line in the vicinity can be used; often the company will permit tapping the line when it is explained the circuit is energized for only a fraction of a second. The first impulse created by gauge indicator contacting either high or low alarm—closes the secondary, holding circuit at the panel board, which is equipped with a locking arrangement. The secondary circuit energizes the alarm and the primary or circuit between the gauge and panel board is cut out entirely from the system.

Both primary and secondary circuits are 110 volt. Line drop between gauge and panel board is of no concern as our special double action relay is designed to function perfectly at a drop of 33⅓%.

For Waterworks

The Gauge shown in the illustration is connected by means of a small pipe to the outlet of the stand pipe, reservoir or tank. When the water pressure is turned into the Gauge the dial hand moves in the same manner as a pressure gauge, being affected by the height of water in the stand pipe or tank.

You will notice two extra hands on the dial of the gauge. These are the electrical contacts. They are movable and may be set where desired. Suppose your stand pipe is 50 feet high and you desire to pump when the water gets low, say 10 feet deep, the dial hand will assume a 10-foot location and then the low movable hand is set to touch it. When you pump your stand pipe full the dial hand moves up accordingly and the high movable hand is set to touch it. This makes the contact High and Low. After once being set it does not need to be changed. When the tank is full and the dial hand makes a contact with the movable hand a circuit is closed over the wires and electricity energizes the magnet coils, which trip and close the switch. When this happens the circuit between the Gauge and the panel board is opened so no short circuit will exist.

To stop the bell ringing a knife switch is opened. Later this switch may be closed and the relay switch reset by merely lifting the little lever and the mechanism will repeat the operation when the water or pressure in the standpipe or tank is low or high.

The Alarm Gauge we will furnish for any height of standpipe, tank or reservoir; simply tell us how high it is. Or, we will furnish a gauge for any pressure up to 300 pounds per square inch. The electrical contact hands are movable and can be set by the purchaser where desired.

This outfit is intended to be operated by Alternating Current, taken from any wire in the pumping station or office; either 110 or 220 volts may be used.

Complete Ready to Install Only $55

This electric alarm outfit when properly installed (they are very simple to install and connect up) will not be affected by this surge of pressure or the pulsation of pumps and will save time, labor and operating because they signal exactly when the pumping should be stopped or started.

The maintenance expense is practically nothing. The electrical consumption is practically nothing or at the most only a few cents a month. No labor or attention is required to keep it in good condition; with just reasonable care it should serve a lifetime. In very cold climates the contact pressure gauge should be protected from freezing by either housing and packing it in sawdust or by placing it in a pit close to the outlet pipe. Especially where stand pipes or reservoirs are located some considerable distance, perhaps several hundred feet or several miles from the pumping station, this outfit will save worry and apprehension on the part of the man in charge of the pumping station, as it will signal when work should be started and pumping should be stopped.

No. E407. Long Distance Alarm. Complete as described and illustrated with large 5-inch bell or loud ringing gong for 110-volt line or lamp socket.

Only.................... **$55.00**

Special Waterworks Recording Pressure Gauges

This instrument is special for the purpose intended and very reasonable in price for the first class movement and workmanship. The pressure springs are unusually large and the clock movements are enclosed in dustproof cases.

In competition with all other makes this Recording Gauge was selected and adopted by the Waterworks Engineers of Chicago and made the STANDARD for the department, and the City now has over 95 in use.

The charts cover practically all ranges of pressure, and we can supply the proper gauge and the right charts for your pressure conditions.

No Complicated Mechanism

Their Simplicity and Ruggedness, in all kinds of usage, are surprising. Without repairs and but little attention they run year after year. There is no particular skill required to install these gauges, as they are connected just as an ordinary pressure gauge. The attention required is but slight—the clock movement is to be wound up, the charts to be changed and the recording pen filled with ink.

On account of the design of the cases and the ringless OPEN FACE features, it is as clear and easy to read as other gauges with charts of ten and twelve inch diameters. Instrument No. R885, priced at $47.50 net, is a favorite with hundreds of Superintendents.

Keeping close check on every variation in pressure and influencing your men to regularly attend their duties. If such duties are not attended, the chart shows this inattention and reports to you fully, whenever you may care to look at it. Thus, as if your personal attention had been given, your plant runs regularly and smoothly. The charts are then filed for future reference, and should you care to learn what was going on in your plant, even years afterward, quickly refer to the charts of the DARLEY GRAPHIC RECORDING GAUGES.

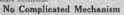

No. R875. Dial. 6¾ inches Diameter, each...	**$37.75**
No. R885. Dial. 8½ inches Diameter, each...	47.50
No. R890. Dial, 10 inches Diameter, each...	62.50

State maximum and minimum pressure with order.

We Are Sole Municipal Distributors of Marsh Recording Gauge Charts—All Sizes

22

Save Time and Money—Look for Them with a Wireless

DARLEY ELECTRIC
Wireless
Pipe Locator and Finder

YOU who may not know Wireless, may place this wonderful instrument in the "quack," bunk or hot-air class. Lots of people have made this mistake, so we don't hold it against any man who hasn't investigated Wireless. We know that such folks feel about Wireless as the man did who saw a giraffe for the first time. "H—l," he exclaimed, "there ain't no such animal!"

Wireless is a revolutionary way to find buried pipes, shutoffs and mains. It is a true wireless instrument, utilizing the same kind of electric waves as the outfits that are found today on nearly every ship afloat. Today any superintendent who has one of these outfits doesn't waste a lot of time having his men dig up the landscape looking for buried pipes, which many times don't appear on the surveys, and waste hundreds of dollars and many hours in this way. He takes his Wireless out on the job and "sics" him on the trail of the missing pipe. And the Wireless gets its electric ear to the ground and hollers in a hoarse buzz until the pipe is right under the man who is looking for it. Then it shuts up.

For those who have not heard before of Wireless let us say that we have made and sold thousands of these instruments. We have shipped them all over the world. We make them so rugged that we ship them half around the world in a fiber box without injury. We make them so simple that any man on earth, with no previous instruction, by following our simple instructions, can find buried pipes and boxes first crack out of the box.

Wireless is built on the following principle: A wire is run from the instrument to a faucet or pipe in one house, and another to an outside pipe, hydrant or faucet in the house next door. This completes a circuit over which the wireless waves set up in the instrument are transmitted. They follow the pipes into the ground and are given off by them in a manner peculiar to such waves. Our receiving outfit picks up these waves and they are plainly heard in the receivers worn by the operator. Due to the construction of this receiving outfit the moment the operator is directly over the pipe THE SOUND STOPS. That is the place to dig, or the line to follow, as the case may be. Wireless will follow a pipe for hundreds of yards, up hill and down, under rivers, under cellar floors, through a network of other pipes, and it never misses. As a pipe locator it is the best little mechanical and electrical hound dawg the world ever saw.

The sound made by pipes followed with wireless is a buzz, like the busy signal in the telephone. It is loud and clear, and absolutely unmistakable. With the receiving apparatus in your hand you can trace the "dead line" of a pipe (the line where there is no buzz from the Wireless) as easily as if the line of the buried pipe were staked out on the ground.

We wish we might publish all of the cases we know of where Wireless has demonstrated its ability, but to do so would fill a newspaper chock full. We have hundreds of letters on file from superintendents all over the world who have found Wireless all and more than we claimed for it.

The practical applications of Wireless are many. It will locate mains in streets, irrespective of pavement or car track, work in winter or summer, or under water. It will locate any service from the main clear into the meter. It locates intersections. Finds yard connections around the plant, on air, gas, water, oil or any other kind of pipe line. It works as well on empty as full pipes. It finds piping in walls, under floors or between floors. It finds bypasses and locates shut-offs.

You will find it the best investment you ever made, and will wonder how you ever got along without it.

SHIPPED ON 30 DAYS FREE TRIAL AND APPROVAL

Wireless Type F $85.00
Forwarded to any part of the World with Express or Forwarding charges prepaid. Custom charges in Canada and other countries to be paid by purchaser.

Same instrument with new Super Sensitive Amplifier Exploring Set, for use under heavy traffic conditions in city business districts, complete $95.00

Trade In Your Old Pipe Locator for $25
If you have an old Pipe Locator, any make, any kind, perhaps 5, 10 or 15 years old, you can trade it in for a new, latest model, Wireless Pipe Locator. Your may be partly or all shot, through wear, tear, age or accident, but we will take it "as is," no questions asked, and make you a liberal allowance of $25.00 on a trade-in.

The new Type F Wireless is $85.00—or only $60.00 if you take advantage of this liberal offer. Return your old instrument by express, charges collect of us, and we will prepay express on your new instrument.

The New Amplifier Coil Pleased This Supt.

New Super Sensitive Amplifier Exploring Set for Wireless Pipe Locator

Present day traffic conditions, especially in business districts, make it desirable to greatly increase the intensity of signals produced through the old style coil. Outside disturbances, such as traffic noises, often make it necessary to delay exploring work until the quiet hours after midnight. Such delays are always expensive, particularly when you are looking for a leaking pipe line.

The new coil was perfected after 3 years of research by our engineers, who experimented with over 100 different types. Their efforts resulted in an exploring coil as perfect as human skill and scientific engineering can make it.

Brings in signals 400% stronger than old coil. Will more than double efficiency of your Pipe Locator.

Trade In Your Old Coil
As a special introductory offer we will accept your old exploring coil and receivers for liberal trade-in allowance of $15.00. We will ship you the wonderful new Exploring Set for $27.50 and your old coil and receivers.

30-Day Free Trial
You are welcome to order the new Exploring Set on our established 30-Day Free Trial and Approval terms. Try it, test it, use it and if it's not all we say, return it at our expense for full credit.

No. D501. New Super Sensitive Amplifier Exploring Set for Wireless Pipe Locator, complete with 2 specially wound receivers, only............................. $37.50

3% OFF All Prices for Cash

Gives 400 Per Cent LOUDER SIGNALS

Shipped on 30 Days' Free Trial and Approval. You to be the sole judge of whether you keep it or not.

Useful Curb Box Cleaner

This new tool will appeal to every water and gas man who has had to clean or clear out old curb boxes that have become so filled with rubbish that the shut-off key cannot be placed or the cock turned.

In some cities it is the practice to dig up the whole box to get at the curb cock because of not having any kind of a tool to clean it out. With this, it is easy, simple and quick to do. You can send for the tool for a 30 days' free trial any time and convince yourself.

It takes any kind of dirt, and the harder it is packed in the better the tool works. After covers have been broken or gone entirely, so the box is full to the ground level, or kids have filled it with stones, pebbles and sand, etc., this tool pulls it all out, digs and lifts it right out.

The digging spades, either OPEN or CLOSED, form a cylindrical PARALLELOGRAM, and the inventor patented the idea. It is really surprising how simple and yet how cleverly he worked out the principle so that it is just exactly right for cleaning curb boxes.

☞ The user just drops it in the curb box and the spades dig into, bite and grip and the load comes up.

☞ Because of the parallelogram action it is the ONLY cleaner that slides evenly parallel with the sides of the box and gets around broken brick, large stones, etc., and brings them up.

Very strong construction, solid, and will stand hard usage.

Well worth many times the price to any Water Dept. or Gas Co. ☞ Try one for 30 days and if it isn't send it back to us.

Made in two sizes, both 6 feet in length.

No. C746. New Curb Box Cleaner for boxes 2 in. diam. and smaller ... **$8.25**

No. C747. New Curb Box Cleaner for boxes 3 in. diam. and larger, including all gate valve boxes ... **$11.75**

☞ Try either size or, better yet, let us send you the set of two ON APPROVAL.

RAWHIDE LEATHER SEAT WASHERS FOR HYDRANT PLUNGERS

Best Quality Leak-Proof

We are prepared to furnish washers for all makes of hydrants. Specify dimensions and we will quote lowest prices.

Rawhide Hydraulic Packing

Rawhide Packing has no equal for Waterworks purposes on centrifugal or reciprocating pumps, stuffing boxes, hydraulic valves, etc. Never becomes hard or glazed. When wet it is as slippery as an eel. It will outlast, by years, any other stuffing box packing. It is made from strips of rawhide, braided, soft and pliable. Good for cold water of any kind and for steam or hot water above 125 degrees F.

The following table shows the sizes carried in stock and the approximate weight per 100 feet.

Diam.	Wt.	Diam.	Wt.
¼ in.		⅝ in.	

No. C318. Rawhide Packing, any size, any quantity, per lb. ... **$3.50**

Trench Pump Diaphragms

Made of best material and fabric for hard use with Rumsey, Cameron, Laad and Gould pumps.

	Outside Diam.	Each
No. C352	11 in.	$1.90
No. C353	12¾ in.	2.60
No. C354	14¾ in.	2.85

Diaphragm Suction Pumps

These pumps are unexcelled for rough and ready service such as required in pumping out excavations, trenches, sewers, cellars and ditches.

Made special with suction hose attachment at the side. This is the standard form used by hundreds of our municipal customers. Threads cut for commercial hose couplings.

Diaphragm has quality of rubber guaranteed to stand hard service. Diaphragm can be replaced by removing only four nuts and lifting off top section. These nuts are brass to prevent rusting.

Waterways are large, permitting an easy flow of liquid.

Hand lever wrought iron, reversible, enabling it to be used vertically or horizontally from either side or the back of the pump. Hand lever is furnished.

We recommend the use of foot valve or strainer with these pumps.

Number	Capacity Per Hour	Suction Hose	Price
791	1800 Gal.	2½ Inch	$21.75
792	3500 Gal.	3 Inch	27.50

Water Superintendents Testing Gauges

Can be attached to any threaded bibb hydrant or to any plain faucet.

No tools necessary.

Every Waterworks man will consider this Gauge a practical and useful necessity.

Without tools it can be attached instantly to any plain or threaded faucet. It will indicate exactly the water pressure in pounds.

Often, with this gauge, one can tell immediately the reason for a complaint. If a service is freezing, or the piping too small and too long to deliver the quantity of water expected or desired, or if sediment is trapped in the line, the gauge will slowly rise and indicate no loss of pressure; in other words, enough water will pass the obstruction and the accumulated pressure will be "full pressure."

If the service leaks, or if some opening exists, like a defective water closet stock that should be fixed, or if any other wasteful escape of water is going on, the gauge will indicate a loss of pressure and cannot possibly show "full pressure."

Many complaints come from sudden areas of low pressure. This can be proved by testing other services in the same neighborhood and if such is the cause the area and extent can be defined. Areas of low pressure are from stopped up mains, or heavy leaks in mains, or from increase in consumption on mains too small to supply the increase and maintain sufficient pressure.

The usual requirement is a 100-lb. dial, but we carry in stock, if wanted, 150-lb. or 200-lb. dials.

No. R301.	With 100-lb. dial	$5.75
No. R302.	With 150-lb. dial	9.00
No. R303.	With 200-lb. dial	9.25

Price includes the leather case.

It Pays to Seal Meters

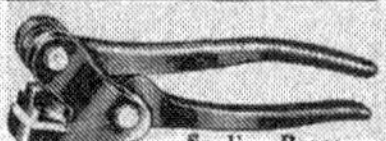

Water, Gas and Electric meters are easily sealed and safeguarded against manipulation and tampering. Thousands of superintendents now take this precaution, at least with active meters. The expense is practically nothing—only a cent or so—and there is everything to gain and nothing to lose by doing it.

Our special Seal Press tool is only 6½ in. long, pocket size, yet the specially designed compound lever makes it the most practical tool of its kind and the price is low.

We will engrave the press to your order with letters which will show on the lead disc, such as "Water Dept." or "Gas Co." or "B. G. & W.," etc. ☞ No extra charge for your engraving.

Wire furnished is flexible stranded copper, tinned and will not corrode. Comes in 100-foot pulls. The meter to be protected is wired, the lead disc, which has two holes through it, is slipped on the wire and squeezed with the Seal Press. Done in a jiffy.

No. 618. Seal Press, engraved, compound lever. Each	$6.20
No. 619. Sealing Wire, copper stranded. Per 100 feet	.80
No. 620. Lead Sealing Discs, perforated. Per 100	.35

Waterworks Meter Reading Book

Our meter reading outfit should be used by every waterworks, having meters to read. Regardless of whether or not you have thousands of meters to read, or less than a hundred, this outfit is best because it will never wear out, it's loose leaf, it's easy to write on and use, it is handy and slips easily into a side pocket and it covers but little.

The holder or binder is stiff, thick, aluminum, made in two parts, riveted and reinforced, with adjustable posts to hold from 200 to 300 slips or consumer records. The cover opens on continuous, smooth hinge and all corners are smooth and rounded so as not to tear clothes or cut your hands. When open, each sheet lays flat, right under the pencil.

The slips or pages are of good tough bond paper, printed with a heading to show Consumer's Name, Street Address, Service No., Meter Size, Name, Number and Dial of Meter. Then, below that, the page is ruled with space for 16 meter readings, with space for DATE, space for the initials of the READER and a column divided into 4 spaces into which to put the figures of the meter READING, then there is another space for REMARKS after each reading. Could not be more complete or handy. The slips are very small indeed, just the right size.

No. R243. Meter Reading Outfit, complete with holder and 200 slips for 200 readings, only ... **$2.85**

No. R244. Extra slips—100, 55c; 500, $2.25; 1,000, $4.00.

Convenience—3,200 Readings in one handy Holder.

Electric Meter Reading Forms—Size 5½ x 8½ Inches

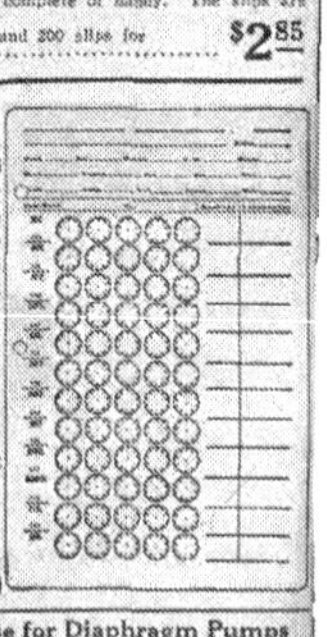

The standard form with spaces for Consumer's Name, Street Address, Folio No., Meter No., Rate, electrical equipment data. Spaces for 12 monthly readings, dates and total monthly consumption.

Our forms are printed on durable white Super Bond paper, 16 lb. substance.

| No. R263. Per 100 sheets | 75c |

No. R264. Binder, slate color canvas cover, metal hinge, holds up to 200 meter sheets ... **$1.40**

Suction Hose for Diaphragm Pumps

2½ in.	No. 909. Length 10 ft.	$14.25
	No. 910. Length 13 ft.	18.79
3 in.	No. 912. Length 10 ft.	$18.54
	No. 913. Length 12 ft.	20.77
	No. 914. Length 13 ft.	24.12

Prices Include Couplings

Electric Leak Locator only $90

A PRACTICAL INSTRUMENT

We are calling it The Electric Leak Locator, because if it were to have a scientific name some one would have to invent it.

It is not a detectaphone or detectograph.

It is not a telephone nor an audion tube.

It is nothing like a phonograph reproducer.

It is not a microphone, yet it more nearly resembles that than anything else. It is an instrument **that hears and greatly magnifies vibrations.**

It consists of a special Microphone-Detector (call it that) mounted on a brass plate with 4 legs and this Microphone-Detector is connected with an Amplifier Battery and Sensitive Wireless Ear Receivers in multiple.

It is very simple to use.

Just set the Microphone-Detector over the main, with the 4 short legs pressed into the earth, if there is no pavement, turn on the switch and listen.

Any leak, or exactly speaking, any leak that needs attention, can be heard if within 20 feet in either direction (a total distance of 40 feet).

If used on paved street the 4 little legs are unscrewed and removed and the plate laid flat on the pavement.

What is the theory? How does it hear the leak?

Escaping water from a main is under pressure. It spurts from the opening and carries air in it and with it.

The leak forces a way through the earth, usually into sewers, drains, etc., and at the opening of escape in the main (at the leak in either words) it forms a pocket into which the air and water discharge under full pressure, churning with it sand, earth and gravel; the velocity of the spurting leak creates vibrations and these vibrations are heard as sound in the Microphone-Detector and in the ear receivers.

The sound, roaring noise we call it, of even a small leak several feet underground is plain, distinct and unmistakable.

A fair sized leak, one about equal to a ⅜ inch round orifice, makes such a roaring noise in this instrument that one inexperienced with it would be inclined to believe the pipe below had actually burst.

It will not detect nor confuse the flow of water through pipes underground, with a leak; it is just purely an underground leak detector or locator.

In actual use the cabinet box, which has a heavy felt edge and a false bottom lined with felt also, is set over the Microphone-Detector, as shown in the illustration where the cabinet is broken away. The reason for this is that wind or any atmospheric disturbance might blow dust or cause blades of grass to rub against the plate.

Any ordinary man can be trusted with this instrument. It is not fragile and no adjustments or calculations are necessary in using it; just push the lever switch on and listen.

It is very light and portable, weighs but 12 pounds complete. No special care necessary in handling it and it can be taken from place to place by hand or in a wagon, street car or auto, without damage or any kind of injury.

Shipped to you on 30 days Free Trial. Express charges prepaid by us.

No. R360 With Battery and Instructions Complete,

$90

In telling about his experience with a street leak a certain Water Superintendent said to us:

"I had a main split or a blown out joint, at the time I did not know which, and found considerable water coming up through the pavement at a street crossing.

"This corner was a low spot at the bottom of two graded streets; a storm sewer carried off considerable of the water that came up to the surface.

"When I arrived at the place my conclusion was that the leak was not under the spot where it showed and probably came from upgrade from one of the lateral branches taken off the main at that place, but I was not exactly sure about it.

"Then I thought of the Electric Leak Locator which I bought some 4 months ago and had never actually tried on a leak but rather expected some day it would prove its usefulness. It seemed to me that here was just the time and place for it to make good and save us a lot of hunting and digging and tearing up asphalt.

"Well, I went and got the instrument and followed one of the laterals uphill, listening as I went along, and in 10 minutes, at a distance of about 150 feet from where it came up through the pavement, I heard it, plain enough! Found it!

"It made good with me! Saved a lot of time, hard work and by noon we had the gates open again and you would never have known anything had happened there, except for the small amount of replaced pavement which we dug through over the leak."

Nearly every Waterworks Superintendent who has and uses our Electric Leak Locator and Finder could tell of similar experiences and how in many cases the instrument probably, and even saved all its cost, the first time used.

Underground leakage and breaks occur with every Waterworks, due to the settlement of trench filling, improper calking of joints, contraction, frost, electrolysis, etc.

When or where the trouble will come cannot be foretold—it is simply up to the Superintendent to find it when it happens.

Our instrument will always repay its small cost.

What Small Leaks Mean
At Average Water Pressure

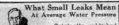

Originally published by the Engineers for the City Waterworks of Cincinnati, Ohio.

SIZE → A leak this size will waste 62,000 gallons a year.

OF → A leak this size will waste 354,000 gallons a year.

LEAK → A leak this size will waste 1,314,000 gallons a year.

Common Causes for Service Leaks

Soft corroded and rust eaten black iron pipe, about twenty-five years old.

Defective curb cock on lead pipe service. Caused by difficulty in placing key to make a shut-off. Result of rough treatment if dirt has partly filled the curb box, which is usually the case.

Broken off iron pipe service. More frequently found at elbows and tees than elsewhere.

Lead service damaged by freezing. Pipe often twisted and expanded and burst open by ice pressure.

Blown Out Cast Iron Lead Joints

Because trenches settle from the weight of cast iron pipe and because lead joints are soft, this kind of a leak is common and most expensive. Often this is the beginning of a "bursted main" that rips up business streets and floods a district.

Unsuspected, under a busy street, they seldom find a way through pavement to the surface, but drain off under ground or run into storm sewers.

Heavy truck, bus and street car traffic on pavement is another cause for this type of leak; also railroad trains, if the main is laid under a railway crossing or run parallel with the tracks, vibration shakes the joint until the soft lead leaks gradually, more and more. Just a 16-inch opening, not much in a 16-inch or larger cast iron main, will flow at least 3½ gallons per hour (OVER 2 MILLION A YEAR) and waste $486.00 annually at 40 lbs. pressure.

Iron service attacked by electrolysis. Stray currents destroy underground pipe very rapidly.

Defective wiped joint due to poor workmanship, a very frequent cause of leaks.

25

A PIPE CUTTER FOR BOTH CAST IRON AND STEEL PIPE

This is the most valuable Pipe Cutter on the market for any Waterworks or Gas Company. It cuts all kinds of pipe, making it a universal tool. It can be used in the shop or in the ditch. Cutting pipe in a ditch with an Elim is the quickest and safest means of cutting it just where and when you want it cut. It entirely answers that purpose.

Every point in contact with the pipe is a cutting disc. You will therefore readily see that the cutter when adjusted on the pipe need only be moved in a small part of a circle in order to cut entirely around.

ONE CUT SAVES THE PRICE

One cut in a ditch partly filled with water—one cut in a tight corner without disturbing other pipes or street work, a cut by divers under water—a hurry-up repair job, cutting out a small section and setting in a new one. Many water Superintendents have reported the cutter to have saved its cost on just ONE such job. You may never use yours for any other than ordinary cutting of pipe, but it will be a most valuable tool should your emergency ever arise, as it has with numerous others.

DESCRIPTION

To place in position in the trench slip out one of the thumb bolts, run the links under the pipe and lock back together—a moment's work. The handle can be lowered on the rod. This allows operation in a narrow ditch.

In cutting pipe in trench all the space required underneath the pipe is enough to place the links carrying the cutter wheels. In making ordinary repairs or inserting specials in a line of pipe, the work can be completed in less time than it takes to dig a hole large enough to use hammer and hardy, all cuts are clean and smooth, without danger of breaking the pipe

or disturbing the joints.

On cast iron pipe the wheels do not cut through, but when a groove has been cut around the pipe to a reasonable depth a little extra pressure on the screw and handle will cause the pipe to crack off. On a 12-inch water main this depth of cut will not exceed one-eighth inch.

TRIAL

This tool is now in use by over 2500 Water Departments and we would be glad to send you one for a try out, you to be the sole and absolute judge of whether you keep it or not.

It is made in two sizes, as listed below and either size will cut either steel or cast iron pipe up to the limit of its capacity.

No. 866. Cuts all kinds of Pipe 4 to 8 inches.
Each.................................**$50.00**

No. 867. Cuts all kinds of Pipe 4 to 12 inches.
Each.................................**$65.00**

No. 868. Extra Cutter Wheels. Per dozen.............................**$7.20**

Kerosene and Gasoline Furnaces
Fast Heating Heavy Duty Type for Water Depts.

Melts 58 lbs. of solder in 16 min. Generator produces a large volume of flame which consumes all carbon. Orifice cleared of foreign particles by turning orifice scraper while furnace is burning. Generator removed through top-plate gate without dismantling furnace. Powerful pump, steel tank, welded bottom, brazed fittings, and shock ring. Removable windshield.

Height, 13 in.; diam., 8 in.; 1 gal. capacity.

No. D991 is fitted with light plumbers top plate. For 4, and 6 in. solder pots. Handle and shield lock for lowering with hook. Wght. of furnace, 12 lbs.

No. D992 is widely used. Heavy top plate for hard service. For 6, 7, and 8 in. solder pots. Wght. of furnace, 12 lbs.

No. D993 is extra heavy with slightly larger top plate. For 6, 7, and 8 in. solder pots. Wght. of furnace, 13 lbs.

Figure 8 Generator and Orifice Scraper

No. D991. Gasoline Furnace, each............**$12.25**
No. D992. Gasoline Furnace, each............13.00
No. D993. Gasoline Furnace, each............14.95
If wanted for burning kerosene, add $1.40 to above prices.

CALKING TOOLS Water or Gas Main Calking Set
Cutting Chisel

No. 8
No. 9 Hammer

No. 3650—Calking Set

1 Regular pattern hand calking iron, ¾-inch thick at point (3)...Each $1.10
1 Regular pattern hand calking iron, ⅝-inch thick at point (5)...Each 1.10
1 ⅝-inch cold chisel (6)..Each 1.10
1 Lead cutting chisel, 3 inches wide at point (7)...............Each 1.20
1 Pipe cutting chisel, with handle (8).........................Each 3.30
1 3½-pound calking hammer, with handle (9).....................Each 3.20
1 Regular pattern yarning iron, ¾-inch thick at point (10)....Each 1.10
Set complete as illustrated...................................Per set 11.95
Leather carrying bag for complete calking set.................Each 10.95

MAIN CLAMPS AND PLUGS

This is for use in testing mains, temporarily closing mains for the night, on account of fire, or for closing mains to branch openings. Or to be used in any emergency where you wish to plug the pipe for a short time.

It can be used to close the mains temporarily but the clamp can remain on as long as desired.

It is made of crucible steel and will easily hold a pressure of 200 pounds or more.

Description To place in position on pipe insert the plug with gasket then hold the main screw against plug while the chain is slipped around the pipe and locked into chain hook. A few turns on the chain hook but brings the arms into position, a few turns by hand of the main screw sets the clamp ready to be tightened with the wrench, which accompanies same. We tap the large plugs with one-half inch pipe plug. This is to allow escape of air, releasing water or connecting gauge, etc.

We have always sent the clamps out subject to approval and now that we are constantly receiving "repeat" orders from cities where they have been in use for some time, we are more than willing that you try same entirely at our risk.

You may not need all the plugs to set, therefore we quote them separately.

No. 1. Main Clamp, complete, less plugs $45.00
No. 2. Main Clamp, complete, less plugs 75.00
No. 3. Main clamp, complete, less plugs 85.00

Plugs for No. 1	Plugs for No. 2	Plugs for No. 3
4 inch......$2.25	12 inch.....$11.50	18 inch.....$25.00
6 inch...... 3.50	14 inch..... 13.00	20 inch..... 30.00
8 inch..... 5.00	16 inch..... 20.00	24 inch..... 35.00
10 inch...... 7.75		

☞ Gaskets included with all plugs.

Test Plugs
for WATER MAINS

The Water Test Plug is a vast improvement over any old style. It is calked in the same way as any other fitting, and to remove it is only necessary to place a crow bar between the lugs, when it can be unscrewed the same as a threaded fitting.
☞ These Water Plugs are for cast iron pipe.
Order No. 548 and give size and kind wanted.

Size	Price	Size	Price
3 in.	$2.50	14 in.	$10.50
4 in.	2.90	16 in.	12.60
6 in.	3.75	18 in.	18.00
8 in.	5.40	20 in.	24.00
10 in.	6.80	24 in.	31.50
12 in.	8.50		

Kerosene Lead Melting Furnace
Equipped with 100 lb. Melting Pot

Melts 100 lbs. of lead in 15 minutes Built for rough use and handling Melting pot made of pressed steel Furnished with a two-gallon seamless steel oil tank with all fittings welded, treated with a special compound, so that it CANNOT RUST. Equipped with extra large air pump, pressure gauge and air release. A special wind shield protects the burners and makes it particularly suitable for out-door use. The Burner, which is the torch type, is made entirely of bronze.

Can also be used for melting babbitt, tin, solder, white or type metal, insulating compound, wax, heating water, etc. Shipping weight 45 lbs.

No. B513. Lead Melting Furnace. Cap'y 100 lbs. Complete **$35.00**

26

Wonderful Leak Detector ☞ Over 12,000 Sold

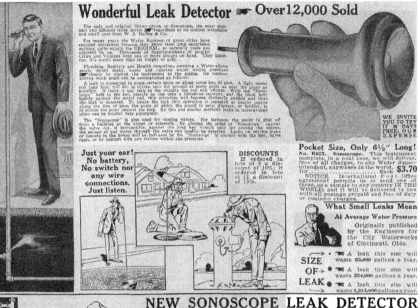

The only and original Water-phone or Sonoscope, the most compact and efficient little device ☞ regardless of its utmost simplicity and small cost from W. S. Darley & Co.

For many years the Water Bureaus of great cities have resisted imitations because they know from long experience nothing quite equals the ORIGINAL, so carefully made and adjusted by us. Thousands of Superintendents of smaller Cities and Villages keep one or more always on hand. They know, too, it's worth more than its weight in gold.

Plumbing, Sanitary and Health inspectors carrying a Water-phone easily detect leaks, waste and running water within premises ☞ simply by placing the instrument to the piping. On outdoor survey work much can be accomplished as follows:

A leak is suspected in some certain main or along some line of pipe. A light metal rod (any kind will do) is driven into the ground at some point as near the pipes as possible. If there is any leak in the vicinity the rod will vibrate. With the "Sonoscope" held to the ear exactly as one uses a telephone receiver, and the point held lightly against the metal rod, this vibration will become distinctly audible and thus the leak is detected. To locate the leak this operation is repeated at nearby points along the line of pipe, the point at which the sound is most distinct, or loudest, is of course the point nearest the leak. By this and similar methods leaks in underground pipes can be located very accurately.

The "Sonoscope" is also used for testing valves. For instance, the water is shut off from a building at the street or sidewalk. By placing the point of "Sonoscope" against the valve (or, if inaccessible, against the long key wrench used for closing the valve), the escape of any water through the valve can readily be detected. Leaks on service pipes or faucets in the home will be indicated by the "Sonoscope" in contact with the key, valve, open, or by contact with any fixture within the premises.

WE INVITE YOU TO TRY IT 30 DAYS FREE! OUR EXPENSE

Just your ear!
No battery,
No switch nor
any wire
connections.
Just listen.

DISCOUNTS
If ordered in lots of 6 a discount of 10%; if ordered in lots of 12, a discount of 15%.

Pocket Size, Only 6½″ Long!
No. R657, Sonoscope. This instrument complete, in a neat case, we will deliver, free of all charges, to any Water Superintendent, anywhere in the world,
for .... Each **$3.70**
NOTICE. International Post Office agreement permits us to send one of these, as a sample to any country IN THE WORLD, and it will be delivered to you with all postage prepaid and free of duty or customs charges.

What Small Leaks Mean
At Average Water Pressure

Originally published by the Engineers for the City Waterworks of Cincinnati, Ohio.

SIZE
OF ›
LEAK

- ☞ A leak this size will waste 62,000 gallons a year.
- ☞ A leak this size will waste 334,000 gallons a year.
- ☞ A leak this size will waste 1,314,000 gallons a year.

NEW SONOSCOPE LEAK DETECTOR
EXTENSION CONTACT ROD
For Hard-To-Get-At Pipes
Designed For Greater Convenience

A new instrument designed for greater convenience and efficiency. Basically the same as the R657 Sonoscope and made to the same scientific precision standards.
The contact rod is steel, cadmium plated rustproof, made in three detachable sections.
Just the thing for pipes and valves located in inaccessible places. The extension rod transmits the vibrations of a leak equally well with one, two or three sections of the rod. The new instrument is highly efficient both with short and long contact rod.
Try the new Sonoscope for 30 Days Free at our expense.
No. R658. New Sonoscope, complete with three section extension contact rod. Including case, shipped prepaid, only...................... **$3.90**

DIPPING NEEDLE SERVICE DEPARTMENT

Dip Needles are magnetized, and hold electricity like a storage battery. We use Tungsten Steel only in Darley Dipping Needles, worth about three times its weight in gold when made into the shape of a needle, as you see it in your instrument.
But like a storage battery, it won't hold forever the electricity stored in it and needles should be recharged every so often.
Once a year—to give your instrument an even break, it should be sent in to us for anything it may need in the way of attention.
To recharge, balance, adjust, inspect and clean up—a minimum charge of only $1.30 is made; not much. If it has been used hard, or met an accident, or been dropped, or if it needs new parts, such as a new chain, pin, bezels, glass, case, or even a new Tungsten Needle and new jeweled bearings, we will make only a nominal charge and GUARANTEE YOU HEREWITH that the outside cost will not be more than $6.00, no matter what it needs nor how much more than that it may cost us to do the job.
☞ Just mail it by Parcel Post, and we'll do the rest.

LEAK LOCATOR SERVICE

To our thousands of customers using this instrument, we offer SERVICE in the way of inspection, repairs and new parts.
To make any old Leak Locator as good as new, the charge is just nominal from $5 to $30, regardless of how many years it has seen service.
☞ Send it in by express and we will take care of the job. Just say Make Repairs and put your name on a paper inside the instrument.

PROMPT WIRELESS REPAIRS AND SERVICE

If you own an instrument of this kind, either ours or any other make, send it to us for repairs or new parts.
We can put in new platinum points, vibrator springs, adjusting screws and furnish new ear receivers, cords and exploring coils.
Even if it has been battered up from 10 or 15 years of service we can restore it to work as good as when new.
Charges are nominal, usually from $7 up, but not over $32.00, even if it costs up more to do the work.
Just say Make Repairs and put your name inside and send by express. No need to write a letter, as we will understand and get busy.

TUBE CUTTER
With Rollers
Light—Strong—Rapid

The two-roller type, with knife-blade wheel and bevel ground reamer. The most practical type for cutting brass, copper and lead tubes.

Perfect, square end cuts result from the use of rollers. Tubes are not twisted, dented, marred nor torn when cut between the wheel and rollers. All friction is eliminated.

The Sharp Wheels cut with surprising speed. The square end cut is always assured. The cutting wheel holds its edge long after hundreds of cuts have been made.

The thin edge of the Cutter Wheel is protected by a positive stop, prevents it from pressing against the rollers.

No. E336. Cuts ⅛ to ⅝ in. tubing. Only $1.95
No. E337. Cuts ⅛ to 1⅜ in. tubing. Only $2.50
No. E338. Cuts ⅝ to 2⅛ in. tubing. Only 3.95

Same size Wheel supplied for Nos. E336, E337, E338 Tube Cutters, 40c each. Wheel Pins supplied for Nos. E336, E337, E338 Tube Cutters, $1 per doz.

High Pressure Test Plugs

Designed especially for gas and water mains ordinarily constructed of cast iron. Place plug into bell end of pipe so that rubber fits into bore, and turn set-screws up into recess in bell. Set-screws are properly located to fit in caulking recess. Expand rubber by turning rectangular nut with a bar. Plug is fitted with threaded connection used for applying pressure to pipe line. These plugs are guaranteed to hold against all ordinary pressures. For safety sake, do not stand in front of any plug when working with high pressures. In ordering, specify size of pipe, material, whether cast iron, wrought iron, steel, concrete, terra cotta; weight, whether standard, extra heavy, double extra heavy, pipe manufacturer's name, maximum testing pressure desired.

For BELL End of Pipe

Number	Size	Price
C101	2-in.	$ 3.44
C102	3-in.	4.64
C103	4-in.	12.72
C104	6-in.	15.12
C105	8-in.	21.52
C106	10-in.	28.44
C107	12-in.	33.12
C108	14-in.	48.72
C109	16-in.	63.84
C110	18-in.	83.04
C111	20-in.	97.44
C112	24-in.	125.36
C113	30-in.	174.32
C114	36-in.	210.12

For SPIGOT End of Pipe

Number	Size	Price
C115	2-in.	$ 5.52
C116	3-in.	6.32
C117	4-in.	12.36
C118	6-in.	17.00
C119	8-in.	32.20
C120	10-in.	36.96
C121	12-in.	49.60
C122	14-in.	71.04
C123	16-in.	89.44
C124	18-in.	110.08
C125	20-in.	133.12
C126	24-in.	162.08

☞ Give correct information as to size and kind of pipe and we will supply the right Plug.

Service Cleaning Is Easy With This 500 lb. Force Pump

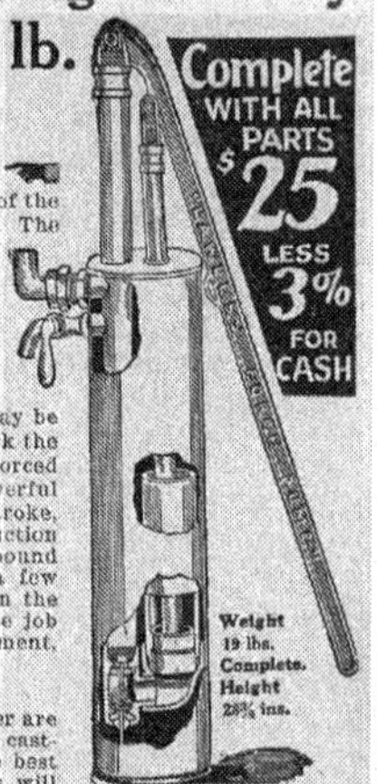
Complete
WITH ALL PARTS
$25
LESS
3%
FOR CASH
Weight
19 lbs.
Complete.
Height
28¾ ins.

☞ **A 19-lb. Pump with a 500-lb. Push** ☜

Connect the quarter bend coupling of the Lawless with the stopped up faucet. The best way is to fasten the pump coupling to one end of a length of lead pipe (often rubber hose will do), to the other end of which is attached another coupling for hitching it up to the service pipe bibb. Fill the pump three-fourths full of water. If the obstruction is far down the pipe, it may be necessary to put more water in. Work the handle. The water in the pump is forced down the service pipe under a powerful pressure, that increases with each stroke, until it exceeds 500 lbs. Any obstruction that can be moved by this force is bound to go back to the main. Usually a few strokes are enough. To test, turn on the pump bibb. If water flows freely, the job is done. No need to tear up the pavement, no need to replace the pipes.

Construction

Outside, inside cylinders and plunger are made of seamless brass tubing. The castings that stand the strain are of the best grade of steam metal. The Lawless will meet the most exacting tests for sturdiness and durability.

Saves Time and Labor Cost

The Lawless Water Main Force Pump has been adopted and is being used with signal success by the water works departments of leading American cities. In every instance it has brought forth unqualified endorsements as an agency for the saving of time and labor costs. It will easily pay for itself on the very first job.

No. 683. Lawless Water Main Force Pump, 500 lb. pressure.... $25.00

W. S. Darley & Co. Offer ☞ 3% Discount for Cash

Tunneling Scoop That Earns Its Cost on the First Job

Used by Gas, Water and Electric men for tunneling under foundations, lawns, walks, roadways, drives, through the roots of trees and shrubs, in hard pan and stony scrabble this scoop tunnels its way—saving time and expense.

This one-piece scoop made of tool steel is light; but heavy enough to punch with. Any length of ¾-inch pipe may be screwed on for a handle. Get into the tough places and out of the tight ones with our Tunneling Scoop.

No. 387. Tunnel Scoop, each $2.25

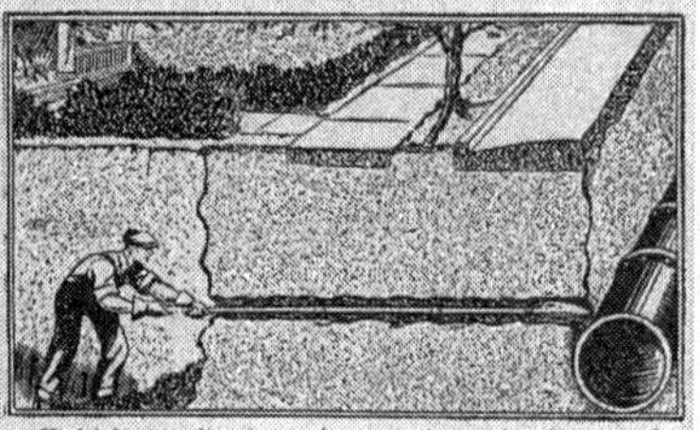

Useful for tunneling for service connections and making way for electrical conduits to signals, posts, signs, etc.

Power Diaphragm Pump Outfits

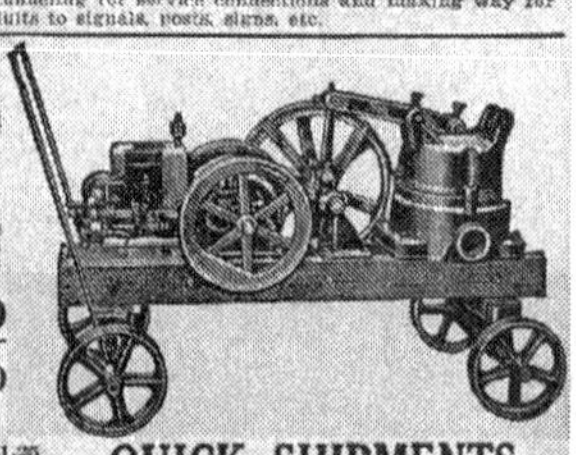

Hundreds of these outfits are giving satisfaction. Complete production of engine, jack and pump in one factory results in a low price that makes this an exceptionally attractive "buy." Compare our price with what others ask.

Construction is the best. Magneto Ignition. Strength and durability are assured by a heavy jack, box type frame carrying a 1¼ in. drop forged steel crankshaft with a 5 in. throw. Bearings are removable die casting 2½ in. long. Malleable iron connecting rod with die cast bearing at large end and removable bronze bushing at small end. The large machine cut gear is 21.6 in. in diameter, 1¾ in. face, and is equipped with guard. Power lever or rocker arm is attached to pump by large hardened and ground pins and removable phosphor bronze bushings form the bearings for these pins. Will operate for hours with little or no attention. Engine 1¾ horse power, pump 3 inch, capacity 4,000 gallons per hour. Handles any kind of water. Will lift 20 feet. Weight of outfit on wheels, as illustrated, 600 lbs.

No. E341. Power Diaphragm Pump, ready to run........ $133.00
No. E342. Engine 1¾ H.P. Capacity 8,500 G.P.H. Takes 4 inch suction hose. Shipping weight, 690 lbs........ $158.00

QUICK SHIPMENTS

Suction Hose—Prices include coupling to pump and a strainer

Strong, reinforced with spiral wire in the thick and tough fabric jackets. Won't kink. Rubber prices are for hose, strainer and coupling, complete.

3 Inch Hose	4 Inch Hose
No. 912. 10 ft. long .. $18.54	No. 912A. 10 ft. long. $31.25
No. 914. 15 ft. long .. 24.12	No. 914A. 15 ft. long. 42.75
No. 915. 20 ft. long .. 29.70	No. 915A. 20 ft. long. 52.50

FOR CUTTING and SWEDGING ¾ and 1 INCH SOFT COPPER WATER TUBING GUARANTEES A PERFECT JOINT

Includes
1½" Tubing Cutter
2 Extra Wheels
2 Tube Holders
2 Forming Tools
2 Clamps
Swedging Tool
Heavy Carrying Case

It makes the square right angle cut so necessary for proper flare and bevel.

It is unnecessary to file or ream the tube before swedging.

Combination Holder, Forming Tool and Clamp guarantees proper bevel—and actually sizes the tubing while being swedged.

No. D779 $22.70

Unbreakable Service and Meter Box Keys

541
C541
Combination Meter and Service Box Key No. D918

The key you've been looking for. Not cast iron. Will not break in service. Made of malleable iron, extra strong. Fits five corner bolt heads on service boxes and meter boxes. For No. C541 size the bolt head measures 1 in. from flat side to opposite corner. For No. 541 size the bolt head measures ⅞ in. from flat side to opposite corner.

No. 541. Service Box Key, each. 25c
No. C541. Meter Box Key, each. . 30c
No. D918. Combination Key for Service and Meter Box, each. . 60c

Curb Box Repair Lids

Fits all 2½ in. Buffalo type Service Boxes, and not only repairs but makes the Service Box better than new.

Screw Bolt **of standard** five corner shape is of forged manganese bronze (not of cast brass) and has a tensile strength of 60,000 pounds. No parts to rust causing breakage. No screws, bolts or rivets used in its assembly.

Operation:—One turn of screw counterclockwise unlocks.

No. 264. Repair Lids, each. 65c

Darley Worm Lock Covers and Lids

Worm lock raises lid as meter is unlocked—lid fits in any position—will not drop in place until it is locked. Bronze washers separate worm and lid and allow easy operation under all conditions. For use on bodies of 18-in. and 20-in. inside diameter. Diameter of lid, 18 in.

Height of cover throat, 4 in.
No. C148. Darley Worm Lock Cover and Lid. $3.95

METER BRUSHES

It is just right for cleaning out scale, etc., from disc chambers and the inside of meter casings. Notice how the all STEEL tufts of bristles are serrated and staggered for the job, with a projecting center and six surrounding rows ☛ in all 138 bunches or tufts of steel bristles set securely in a hardwood block base.

The block or base has a lead core so it can easily be attached to the spindle or shaft of any electric grinder.

Many water superintendents find it DOES SO MUCH and COSTS SO LITTLE that they order 6 to 12 at a time. ☛ Just try one in your meter shop and you'll come back for more.

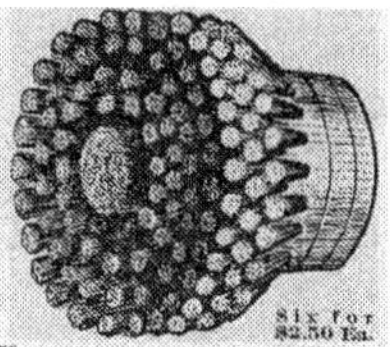

Six for $2.50 Ea.

No. C835. Rotary Meter Brushes, each. $2.90

PORTABLE METER TESTER

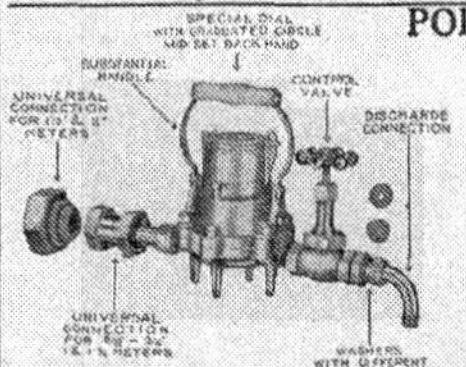

SPECIAL DIAL WITH GRADUATED CIRCLE AND SET BACK HAND
SUBSTANTIAL HANDLE
CONTROL VALVE
UNIVERSAL CONNECTION FOR 1½" & 2" METERS
DISCHARGE CONNECTION
UNIVERSAL CONNECTION FOR ⅝ - ¾ 1 & 1½ METERS
WASHERS WITH DIFFERENT ORIFICES IN THIS UNION

May be used in the shop or the service meters may be tested without removing them from the pipes. The test is made by connecting the testing meter in tandem with the service meter so that the same volume of water passes through both; any error in the service meter is shown by the difference in registration of the two meters.

Complete instructions and a certified test report with each outfit. Weight, 18 lbs.

No. 304. Portable Meter Testing Outfit. $29.50

Water Meter Coupling Yokes

Meter is easily taken out in a few seconds time.

A rapid and convenient means of coupling up meter in meter box. Yoke ells are screwed on riser pipes and frame dropped in place. Turning the handwheel tightens or loosens meter. No tools required. Meter couplings eliminated. With Type No. 2 valve may be used instead of curb cock and service box eliminated. Outlet test cock determines whether meter will register on flow through ⅛ in. orifice. Only four parts—gray iron ells, frame, handwheel, spacer clamp. Gaskets cannot drop out when removing meter. Slide adjustment provides an expansion joint in service line. ☛ When ordering specify whether inlet and outlet are desired with inside iron pipe thread or with flared fittings for copper pipe.

Type No. 1

Type No. 4

	Type No. 1 Meter Yoke—Plain Inlet—Plain Outlet				Type No. 4 Meter Yoke—Valve Inlet—Test Cock Outlet			
Catalog No.	1A	1B	1C	1D	4A	4B	4C	4D
Meter Size	⅝"	⅝"x¾"	¾"	1"	⅝"	⅝"x¾"	¾"	1"
Inlet and outlet inside iron pipe thread.	$1.50	$1.50	$2.16	$2.85	$2.20	$3.20	$3.80	$6.45
Inlet, flared fitting for copper pipe and outlet inside iron pipe thread	1.80	1.80	2.45	3.40	3.40	3.40	4.00	7.00
Inlet and outlet flared fittings for copper pipe	2.15	2.15	2.80	3.95	3.75	3.75	4.35	7.60

Replacement Parts for Nos. 2A, 2, 3, Eclipse Fire Hydrants

Nos. 475-477 Nos. 500-501-502 Nos. 580-581-583 Nos. 490-491

Cast Iron Extension Pieces (Complete with Gaskets and Bolts)
No. 475—To refit No. 2 Hydrant (6 in. long). Each. . . $2.60
For each additional 6 in., add $1.08 to cost.
No. 477—To refit No. 3 Hydrant (6 in. long). Each. . . . 5.40
For each additional 6 in., add $1.46 to cost.
Specify length desired, from 6 in. on. (Operating rod must also be lengthened.)

Iron Stock with Brass Nozzle and Iron Cap (With Gaskets and Bolts)
No. 480—For No. 2 Fire Hydrant, 4-ft. bury with 1-2-in. Nozzle. Each. $15.00
Add or deduct from cost, 80c per ft. of bury.
For two 2 -in. Nozzles, add $2.20 to cost.
For one 2½-in. Nozzle, add 1.40 to cost.
For two 2½-in. Nozzles, add 8.00 to cost.
No. 481—For No. 3 Fire Hydrant, 5-ft. bury, with 1 2½-in. Nozzle. Each 25.60
With 2 2½-in. Nozzles. Each 26.88
For each Nozzle and Cap not required, deduct their respective prices.

Cast Iron Top Stock (Complete with Gaskets and Bolts)
No. 486—For No. 1¼-in. and 1½-in. No. 2A Fire Hydrant. Each. $7.00
No. 487—For No. 2 Fire Hydrant. Each 4.40
If Bolts are not desired, deduct 18c from cost.

Fire Hydrant Handle
No. 490—For 1¼-in. No. 2A Hydrant. $0.48
No. 491—For 1½-in. No. 2A Hydrant. 48

Fire Hydrant Wrench
No. 492—For No. 2 Hydrant. $0.80
No. 493—For No. 3 Hydrant. 80

Bronze Top Screw
No. 500—To refit 1¼ or 1½-in. No. 2A Fire Hydrant. Ea. $2.20
No. 501—To refit No. 2 Fire Hydrant. Each. 2.30
No. 503—To refit No. 3 Fire Hydrant. Each. 4.40

Operating Rod
Always specify depth of bury of hydrant from pavement line to center of inlet.
No. 505A—To refit 1¼ or 1½-in. No. 2A Fire Hydrant, 5-ft. bury. Each $0.80
No. 505B—To refit No. 2 Fire Hydrant, 5-ft. bury. Each. .80
Add or deduct 8c per ft. from above costs for differences in length.
No. 507—To refit No. 3 Fire Hydrant, 5-ft. bury. Each. 1.78
Add or deduct 18c per ft. from above costs for differences in length.

Bronze Top Connecting Nut
No. 510—To refit 1¼ or 1½-in. No. 2A Fire Hydrant. Ea. $2.30
No. 511—To refit No. 2 Fire Hydrant. Each. 2.30
No. 513—To refit No. 3 Fire Hydrant. Each. 4.30

Bronze Seat Ring
No. 520—To refit 1¼ or 1½-in. No. 2A Fire Hydrant. Ea. $2.00
No. 521—To refit No. 2 Fire Hydrant. Each. 2.80
No. 523—To refit No. 3 Fire Hydrant. Each. 3.60

Plunger Complete
No. 530—To refit 1¼ or 1½-in. No. 2A Fire Hydrant. Ea. $1.40
No. 531—To refit No. 2 Fire Hydrant. Each. 5.60
No. 533—To refit No. 3 Fire Hydrant. Each. 7.20

Rubber Seat Washer for Plunger
No. 540—To refit 1¼ or 1½-in. No. 2A Fire Hydrant. Ea. $0.40
No. 541—To refit No. 2 Fire Hydrant. Each. 44
No. 543—To refit No. 3 Fire Hydrant. Each. 1.32

Plunger Cup Washer (Two to Plunger)
No. 560—To refit 1¼ or 1½-in. No. 2A Fire Hydrant. Ea. $0.05
No. 561—To refit No. 2 Fire Hydrant. Each. 05
No. 563—To refit No. 3 Fire Hydrant. Each. 07

Iron Bottom (Complete with Gasket, Bolts and Bronze Waste Tube)
No. 569—To fit 1¼-in. Pipe (No. 2A Hydrant). Each $ 7.50
No. 570—To fit 1½-in. Pipe (No. 2A Hydrant). Each $ 7.40
No. 571—To fit 2 -in. Pipe (No. 2 Hydrant). Each. 9.90
No. 573—To fit 4 -in. Pipe (No. 3 Hydrant). Each. 13.60
No. 575—To fit 6 -in. Pipe (No. 3 Hydrant). Each. 15.20
For Standard Flanged Inlet Connection, Faced and Drilled, add $2.00 to cost.
For Companion Flange, add $2.20 to cost.
Specify Style of Inlet when ordering.

Iron Top Cap (Complete with Gasket and Bolts)
No. 580—To refit 1¼ or 1½-in. No. 2A Fire Hydrant. Ea. $1.72
No. 581—To refit No. 2 Fire Hydrant. Each. 1.72
No. 583—To refit No. 3 Fire Hydrant. Each. 2.88

Iron Nozzle Cap (Complete with Chain and Gasket) (Threaded to Order)
No. 589—To fit 1¼-in. Nozzle. Each. $0.64
No. 590—To fit 1½-in. Nozzle. Each. 80
No. 591—To fit 2 -in. Nozzle. Each. 80
No. 592—To fit 2½-in. Nozzle. Each. 1.60
No. 594—To fit 4½-in. Nozzle. Each. 2.40

Bronze Nozzle for No. 2A and No. 2 Fire Hydrants (Threaded to Order)
No. 599 —To fit 1¼-in. Coupling. Each. $1.60
No. 600 —To fit 1½-in. Coupling. Each. 1.60
No. 601 —To fit 2 -in. Coupling. Each. 1.88
No. 601A—To fit 2½-in. Coupling. Each. 2.00
Furnish Sample of Hose Thread.

Bronze Nozzle for No. 3 Fire Hydrant (Threaded to Order)
No. 602—To fit 2½-in. Coupling. Each $3.40
No. 603—To fit 3½-in. Coupling. Each. 8.60
No. 604—To fit 4½-in. Coupling. Each. 8.60

Bronze Waste Tube (With Leather Gasket)
No. 610—To refit 1¼ or 1½-in. No. 2A Fire Hydrant. Ea. $1.80
No. 611—To refit No. 2 Fire Hydrant. Each. 1.80
No. 614—To refit No. 3 Fire Hydrant. Each. 3.20

Bronze Top Bushing
No. 620—To refit 1¼ or 1½-in. No. 2A Fire Hydrant. Ea. $1.40
No. 621—To refit No. 2 Fire Hydrant. Each. 1.40
No. 624—To refit No. 3 Fire Hydrant. Each. 1.80

Nos. 480-481

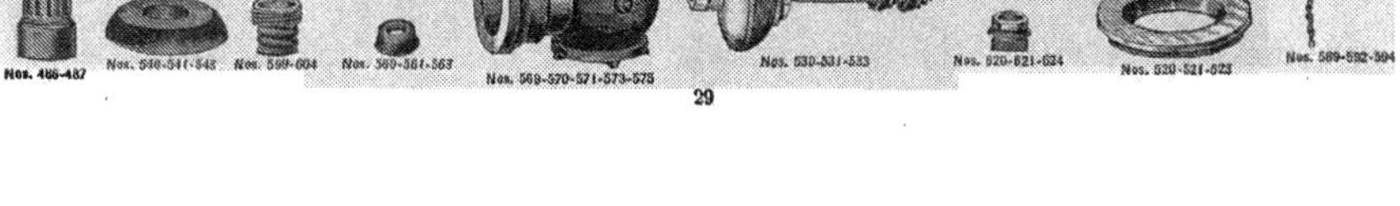

Nos. 486-487 Nos. 482-493 Nos. 540-541-543 Nos. 599-604 Nos. 510-511-513 Nos. 560-561-563 Nos. 569-570-571-573-575 Nos. 530-531-533 Nos. 610-611-614 Nos. 520-521-524 Nos. 505A-505B-507 Nos. 520-521-523 Nos. 589-592-594

Beaver Ratchet Threader

The only ratchet tool made threading 1 inch and all smaller sizes.

The most useful small set one can have. It is not only handy for bench work, and to carry, but will thread pipe in cramped places where a two-handle stock cannot be used. Conceded to be the best ratchet tool of its type ever made. Each outfit includes handy carrier for dies.

No. B931. With ½, ¾, 1" Dies . . $11.25
No. B932. With ⅜, ½, ¾, 1" Dies 13.50
No. B933. With ¼, ⅜, ½, ¾, 1" Dies 15.75
No. B934. With ⅛, ¼, ⅜, ½, ¾, 1" Dies 18.00

A NEW LOW PRICE CLAMP

Stops Bell and Spigot Joint Leaks

A new design which provides a most economical method of permanently repairing bell and spigot leaks in water and gas mains. Construction of clamp and gasket assures an efficient and positive seal that is not only flexible but absolutely leak-proof under all conditions. Made in all standard sizes.

PERMANENT—The rigidity of this clamp and the sturdiness of its construction embodies every possible assurance of permanent repair. It is built for pressures greatly in excess of any that will be encountered. No adjustments are necessary. The clamp is tailored to the joint. ☞ All bolts are steel, rustproof cadmium plated.

EASILY INSTALLED—Simplicity in design minimizes number of parts and eliminates the necessity of adjustments. Fewer parts, especially in the larger sizes, also facilitate and speed up installation.

Cross section showing how gasket is sealed—making the pipe repair absolutely permanent.

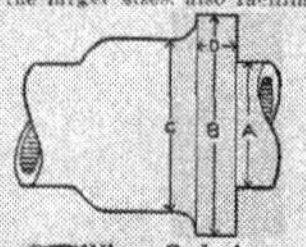

☞ When Ordering give us pipe dimensions A-B-C-D as shown above.

Extremely heavy construction of sections to prevent any possible distortion in drawing up. Hand sledged for even bearing completely surrounding spigot and as a test of perfect annealing.

ECONOMICAL—Performance has proven the economy of this clamp. The ease of application reduces installation cost to a minimum and the permanence of the installation eliminates maintenance expense.

Paranite rubber gaskets, factory sized to avoid cutting and fitting for installation.

Order No. D853 and give us pipe dimensions as shown at left

Cast Iron Pipe Sizes	3"	4"	6"	8"	10"	12"	16"	20"	24"	30"	36"
Weight, lbs.	14½	16	24½	28	38½	46½	63½	91	119	146	195
Price, Each	$3.28	3.57	4.90	8.07	11.40	13.23	19.38	26.16	33.39	49.62	56.78

3%
For Cash With Order or C. O. D.

New Useful Curb Box Cleaner

ALWAYS BRINGS UP A LOAD

This new curb box cleaner is the invention and design of a practical waterworks man. It will appeal to every water and gas man who has had to clean or clear out old curb boxes that have become so filled with rubbish that the shut-off key can not be placed or the cock turned.

In some cities it is the practice to dig up the whole box to get at the curb cock because of not having any kind of a tool to clean it out. With this, it is easy, simple and quick to do. You can send for the tool for a 30 days' free trial any time and convince yourself.

It takes any kind of dirt, and the harder it is packed in the better the tool works. After covers have been broken or gone entirely, so the box is full to the ground level, or kids have filled it with stones, pebbles and sand, etc., this tool pulls it all out, digs and lifts it right out.

It's sort of a clam-shell bucket, works on the same principle. There is an auxiliary rod inside that controls the diggers. By a twist of this handle (on top, see it in picture), the pair of digging spades (at the bottom) open. The user drops it in the curb box and the spades bite into a handful, a turn of the handle and the spades grip the load and up it comes.

Very strong construction, solid, and will stand hard usage. Well worth the price to any water department or gas company. Weight, 12 lbs.; length, 5 ft. 4 in.

No. E625. Curb Box Cleaner **$17.50**

Leak Clamps

Will effectively and lastingly seal pinhole and small leaks in mains or services, either steel or cast iron pipe.

☞ Made of a hot-rolled steel band with two ears of certified malleable iron securely riveted on. In use the band is sprung around the pipe, a piece of rubber packing placed underneath the band and over the leak, and the band tightened down with the bolt.

Band Clamp offers a simple and inexpensive, yet very effectual method of permanently sealing leaks in pipe lines.

Up to 5" bands are 2" wide, larger sizes 3" wide.

☞ Order No. D161 Leak Clamps and give sizes.

☞ **NOTE HOW SMALL THE COST**

Size Pipe	Each	Size Pipe	Each
1" to 2"	$ .65	10"	$1.75
2½" & 3" OD.	.70	11⅝" C.	1.90
3" & 3½"	.75	12"	2.00
4"	.80	14" OD.	2.25
5"	.85	16" OD.	2.50
5⅝" C.	1.20	18" OD.	2.75
6"	1.25	20" OD.	3.00
6⅝" C.	1.30	22" OD.	3.25
7⅝" C.	1.40	24" OD.	3.50
8"	1.55	26" OD.	3.75
9⅝" C.	1.65		

HYDRANT TEST GAUGE

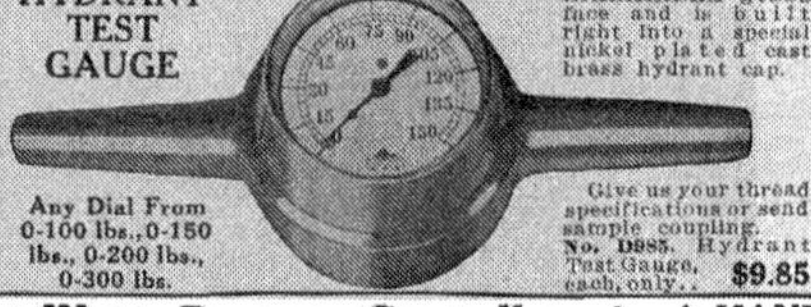

Any Dial From 0-100 lbs., 0-150 lbs., 0-200 lbs., 0-300 lbs.

The gauge has UNBREAKABLE glass face and is built right into a special nickel plated cast brass hydrant cap.

Give us your thread specifications or send sample coupling.

No. D985. Hydrant Test Gauge, each, only . . $9.85

Meyer's Patent Water Pressure Controller

When used on risers, shutting off of branch stop-cocks, blowing out of pipes, waiting for pipes to drain out, or using plugs is avoided.

To repair leak in service pipe, or to put on stop-cock under pressure, just cut the pipe; one stream of water similar to that of a hose will escape straight, as the bore of the pipe is not reduced by a faucet, etc. The stream will be from 1 to 3 feet long. You cannot get wet as the water does not scatter when it leaves the pipe, only when it strikes an object or cellar bottom. You insert the controller into the pipe at once and give wing-nut three turns which stops the flow. You will experience no difficulty in doing this, as the washers are smaller than the bore of the pipe. After having cut the pipe and controlled the pressure (the whole operation having taken only a few seconds), slip a round water-way stop-cock over the stem of the Controller and join to the pipe; then release the wing-nut, pull Controller out through stop-cock and shut stop-cock off, and join other end of pipe to stop-cock. Supplied complete with washers for ½", ⅝", ¾", 1" and 1¼" pipe. Equally useful on copper, brass or other pipe of above sizes.

No. 195 Pressure Controller, complete, each **$4.80**

W. S. Darley & Co. Offer ☞ 3% Discount for Cash

A HANDY PUMP FOR EVERY SUPT.

The pump is held steady by the foot rest. When the handle is worked up and down it pumps the pressure up to 250 lbs. in the pipe and whatever is clogging the service or drain must move. It takes only a few minutes to do the job and the water will flow freely.

In many cases of complaints this little pump will save tearing up pavements, replacing pipes.

The complete outfit includes a heavy seamless brass pump with brass ball valves, two 4 foot lengths of ½ inch hose and a heavy brass conical tip with tapered thread which can be screwed to any size pipe thread from ⅝ to 1 inch. Threaded full length on the outside so that it will easily catch the threads of the pipe. This tip is removable so that you can easily attach a union coupling if you prefer it.

We will be glad to send you one of these pumps subject to your approval.

No. 101. Each $8.85

Special Waterworks DROP FORGED Adjustable $1.75 HYDRANT WRENCH and HOSE SPANNER

Adjustable  *Unbreakable*

For many years we made our Adjustable Waterworks Wrenches of malleable steel. But, with the insistent demand for a stronger, tougher wrench, we are now Drop-Forging our wrench of a high grade CARBON STEEL. The steel we use was selected after exhaustive tests as providing the strongest, toughest and most desirable.

Our Drop Forged Adjustable Hydrant Wrench is **absolutely breakproof**—no matter how hard the work.

Opening in head may be regulated to fit any size pentagon nut up to 1¾-inch across face of same by turning the handle to right or left for size desired.

A Supt. needs a tool that's a help—He wants a wrench on which he can pull his last ounce, if necessary, and know the tool will take all he can give.

No. C285. Drop Forged Adjustable Hydrant Wrench and Spanner, each **$1.75**

IF YOU DON'T SEE WHAT YOU WANT WRITE US

Stronger than the Pipe and So Easy to Install

For Service Leaks, Small Mains, Etc.

These Leak Clamps are for Wrought Iron, Steel, or Lead pipe, such as services, small mains, risers, etc.

Clamps are packed in cartons, twelve to a box, ½ in. to 2 in., and six to a box 2½ in. to 4 in., inclusive.

We especially recommend buying them a box at a time, to have some on hand in case of trouble.

Order No. C745

Size of pipe	½"	¾"	1"	1¼"	1½"
Price, each	$.60	$.68	$.75	$.90	$1.06
Size of pipe	2"	2½"	3"	3½"	4"
Price, each	$1.20	$1.50	$1.88	$2.25	$3.00

HANDY TO HAVE AROUND — ONLY $3.90

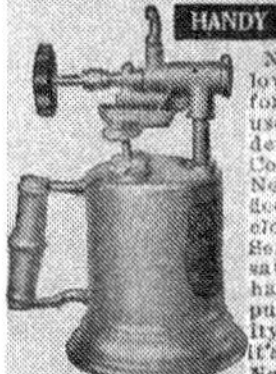

New high-quality low priced blo-torch for non-professional users. Improved design burner develops extremely hot flame. Cool composition valve handle. New design needle prevents orifice enlargement. Completely enclosed trouble-proof pump unit. Seamless brass tank with dull satin finish. Smart red lacquered handle. Top filling of tank by pump removal. One quart capacity. While prices are about half it's a good time to get one.

No. C786. Each, only **$3.90**

STANDARD SPLIT SLEEVES FOR WATERWORKS

Approved by Superintendents everywhere who find they do save many, many times the small cost.

Thousands of dollars are saved annually by Gas and Waterworks, in using our Standard Split Sleeves. It can be driven over a leak instantly and checks the flow to a mere trickle—or when you can reduce the pressure and use our clamp properly with the gasket, it stops the leak tight, thus saving the heavy cost of tearing out a section of a line. As the Emergency Clamp is made of malleable iron, it can be "fitted" to irregular surfaced pipe with a sledge hammer.

For repairing leaks in MAINS or any pipe line, instantly, permanently. The sleeve consists of a malleable iron cylinder, divided in halves, hinged full length along one side and fitted with tightening bolts on the other. A gasket completes the clamp.

When a leak appears, apply and tighten the sleeve. The repair is permanent, and the pipe is stronger than it was in the first place. For a split, use a long strip of packing and two or more sleeves butted together.

Because made of malleable iron they will stand up or last indefinitely under electrolysis: ☞ in the worst soil ☞ or atmospheric conditions ☞ a positive permanent repair for any pressure up to 800 lbs.

The only tool needed to apply Standards is a small wrench. Note long bolts and raised lugs for wrench convenience.

The Standard Sleeve is as much a part of safety first equipment, as fire hose. We carry all sizes in stock.

For CAST IRON Pipe, with heavy ribs, which insure security, guaranteed to stop any leak, GAS, WATER or STEAM, under any pressure up to 800 lbs.

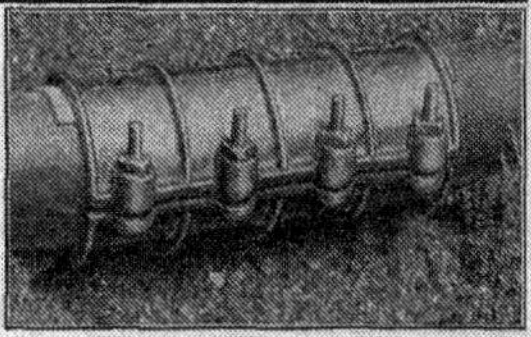

Painted and ready for use. Prices are for clamps complete with packing. Order by Number. ☞

For Cast Iron Mains

No.	Size Pipe	Weight, Each, Lbs.	Price Each
749	4 in.	8½	$ 6.75
750	6 in.	14	10.50
751	8 in.	29	12.00
752	10 in.	45	15.00
753	12 in.	60	18.00

For ☞ Wrought Iron or Steel Mains

Size Pipe	Length	Wt., Lbs.	Price Each
5 in.	7 in.	11	$ 7.50
6 in.	8 in.	14	9.00
7 in.	9 in.	16½	11.00
8 in.	9½ in.	29	12.00
9 in.	10 in.	35	13.50
10 in.	11 in.	45	15.00
12 in.	12 in.	60	18.00

☞ Order these by size.

☞ Our warehouse stock is complete.
☞ **Write or telegraph orders.**

3% Discount FOR CASH WITH ORDER OR C. O. D. SHIPMENTS

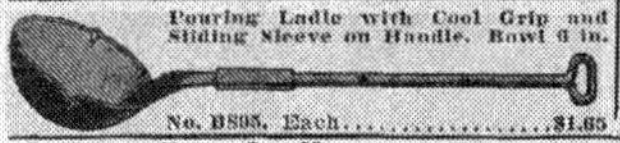

Pouring Ladle with Cool Grip and Sliding Sleeve on Handle. Bowl 6 in.

No. B895. Each $1.65

Pouring Pots Cast Iron

No. B986. Capacity, 50 pounds of lead. Each, **$2.90**

No. B987 Capacity 150 lbs. Each, **$4.50**

Bottom Pour Ladle

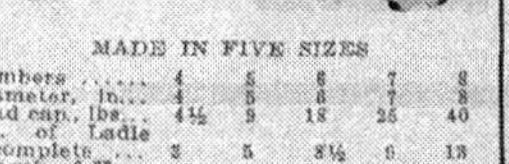
INTERIOR VIEW OF LADLE ☞

Self-skimming. The only ladle accepted by the Underwriter's Lab. The most valuable device ever invented for the pouring of metals. It is the only ladle made that takes the metal from the bottom of the bowl instead of the top, thereby avoiding all the scum and scoria that gathers on the surface and saving the loss of time and material in skimming. The sliding hand grip is another valuable safety feature.

MADE IN FIVE SIZES

Numbers	4	5	6	7	8
Diameter, in.	4	5	6	7	8
Lead cap., lbs.	4½	9	18	25	40
Wt. of Ladle complete	3	5	8½	6	13
Length of Handle, in.	14	16	20	23	30
Price, each	$2.40	$2.70	$3.60	$3.90	$4.20

No. 1398 SPECIAL CUTTING-IN TEES

Easy to Install

These specials are for use when cutting into a street main for the introduction of extra hydrants, new street main, or other large service.

The specials are enlarged back of the bell, thus they can be easily slipped over the pipe and drawn up into position ready for calking. Only two joints are required; there is less labor, less excavating, less time for the water to be turned off and a variation of an inch or two in the length of piece cut out, does not matter.

Tees 6-inch and larger with reducing branches (not smaller than 4-inch) also carried in stock.

TABLE OF SIZES AND WEIGHTS

Cast Iron Pipe Sizes	3"	4"	6"	8"	10"	12"	14"	16"
Length of pipe to cut out	16"	19"	21"	23"	24"	31"	36"	36"
Approximate weight, each, Pounds	140	155	200	350	470	650	800	1200
Price, Each	$17.85	$19.77	$24.00	$42.00	$56.40	$78.00	$96.00	$144.00

Double Length Clamps

No.	Size Pipe	Length	Wt. Lbs.	Price Each
8653	1 in.	8 in.	3	$1.50
8654	1¼ in.	8½ in.	4	2.00
8655	1½ in.	9 in.	5	2.50
8656	2 in.	10 in.	7½	3.00
8657	2½ in.	10½ in.	8	3.75
8658	3 in.	11 in.	10½	4.70
8659	3½ in.	11½ in.	12	5.50
8660	4 in.	12 in.	12½	6.50
8661	4½ in.	12½ in.	17	10.75
8662	5 in.	13 in.	20	12.00
8663	6 in.	15 in.	28	13.50

Same design and construction as the others except that these are about double the length, to take care of long splits or very bad corroded spots. No job at all to put on.

☞ Owing to the long open hinge, the two halves of the clamp may be slid together along the pipe and make it a simple matter to fit when crowded for room.

ASBESTOS LEAD JOINT RUNNERS

Made of specially prepared pure asbestos rope, having a ferrule on each end and provided with a clamping device for fastening in place on the pipe.

No.	Size	For Pipe	Price
115	⅝ in. sq.	2, 3 and 4	$ 1.80
116	¾ in. sq.	4, 5 and 6	2.08
117	1 in. sq.	6, 8 and 10	3.64
118	1 in. sq.	10, 12 and 14	4.68
119	1¼ in. sq.	16, 18 and 20	9.62
120	1¼ in. sq.	 24	10.66
121	1¼ in. sq.	 36	12.22
122	1½ in. sq.	 42	14.17
123	1½ in. sq.	 48	16.38

★ We Are Always Ready to Give You Service ★ Telegraph Day or Night at Our Expense ★

The Best Made Sewer Rods at Any Price

Experience develops that the best rods can only be made from oiled oak wood, and the best coupling is the one that joints together the easiest and quickest and slides freely.

Our Security couplings hold the rods firmly together WITHOUT SLACK and without danger of separation in the sewer pipe and when removed they uncouple easily.

The design is simple and not likely to get broken in a manhole or when thrown about in trucks or autos. This is important because the handling sewer rods get is usually pretty rough and severe.

Couplings are made from genuine MALLEABLE IRON. Tough stuff! Couplings are swedged or shrunk down very tightly on to the rods at the curved places at the ends. Security couplings can not come loose or come off. ☞ No rivets are used in making Security Rods as wood always breaks at rivet holes.

Although our rods hold together without slack and can not possibly come apart in the sewer or duct, they will lend themselves to moderate bends or curves.

They are not like other sewer rods because they are light in weight and being made BELLIED IN THE CENTER the friction between the rod and the pipe is reduced to a minimum, thereby enabling one man to handle long runs easily and q u i c k l y. ☞ When locked or coupled together in sewers our rods will not buckle under pressure nor bind against the side walls.

Made in standard lengths of 3 and 4 feet long. Prompt shipments from our large stocks.

SECURITY RODS

Yes Sir! Made From OILED OAK The Best For Sewer Rods

The COUPLING that will NOT come apart in the sewer.

No. 771. Security Rods, 3 feet long.
Weight only 28 ounces per rod..........Our Price Each **65c**

No. 772. Security Rods, 4 feet long.
Weight only 32 ounces per rod..........Our Price Each **75c**

For Use With Security Rods ☞ A Tool for Every Emergency

Tools that soon repay their small cost

Obstructions in pipes are of different kinds. In order to remove them quickly and thoroughly it is essential that one should have the proper tool. We call particular attention to the Root Cutter No. 8. This does the work of removing roots from sewers so well and economically that we unhesitatingly recommend it as the best tool for the purpose to be obtained. All of the line of tools are practical, durable and inexpensive.

Size of Sewers	4-in.	5-in.	6-in.	8-in.
No. 2. Scrapers	$ 9.00	$ 9.60	$11.20	$12.80
No. 3. Screw	4.80	4.80	4.80	4.80
No. 4. Plunger	7.20	7.20	7.20	7.20
No. 5. Gouge	4.80	4.80	4.80	4.80
No. 6. Brushes	6.40	8.00	9.60	12.80
No. 7. Claw	4.80	4.80	4.80	4.80
No. 8. Root Cutter	14.40	15.35	17.30	19.95

Size of Sewers	10-in.	12-in.	15-in.
No. 2. Scraper	$17.30	$19.20	$22.95
No. 6. Brushes	16.00	18.40	22.40
No. 8. Root Cutter	19.20	24.00	28.80

For a sewer full of sand, or nearly full, use Scraper —Cut No. 2. For paper, rags, etc., use Screw—Cut No. 3; or Claw—Cut No. 7. For unwined joints, with cement protruding, use Gouge—Cut No. 5. For sewers

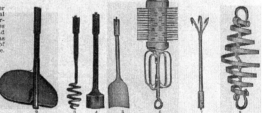

subject to accumulations of grease, use Wire Brush —Cut No. 6 (to be used with a rope). For smoothing round or square for 3-in. or 4-in. conduit). For removing obstructions before drawing

cables, use Plunger—Cut No. 4 (furnished either round or square for 3-in. or 4-in. conduit). For removal of roots the Root Cutter—Cut No. 8

UTILITY GAS MASKS

$12.85

There is not a mask on the market today selling for less than $50.00 that will offer greater and more dependable protection against smoke, organic and acid gases, ammonia, sulphur dioxide, chlorine, industrial gases, and in general all gases except carbon monoxide and hydrocyanic gases.

Small, compact, efficient, with the filters built directly into the gas-tight facepiece. The chemical filter cartridges contain newly developed absorbents, highly concentrated and with higher efficiencies than the chemicals used in canister masks. The cartridge is sealed in place by a screw cap. All inhaled air is drawn through the cartridge, which filters and absorbs all smoke and gas, allowing only pure air to pass through into the wearer's lungs.

Snug, comfortable fit on all faces. Full freedom and use of both hands is allowed the user. No canisters or air lines to strap onto the body, interfering and limiting the capacity of the user.

Wide vision lens are NON-SHATTERABLE glass

and NON-FOGGING as a result of special head piece design.

Easily and quickly slipped on. The rubber headharness requires only one adjustment—no readjustment necessary every time mask is put on.

Cartridges do not deteriorate when not in use. Replacement cartridges easily installed in a few seconds.

No. E104. Utility Gas Mask for smoke and gas, with sturdy composition carrying case, ready for instant use, only...... **$12.85**

Extra Filter Cartridge for smoke, gas and ammonia 98c

Extra Filter Cartridge for chlorine gas......98c

Extra Filter Cartridges for Any Gas

Tell us the conditions under which your mask must serve, the gases you would expect to encounter in your work and if necessary special cartridges will be furnished at no extra charge.

Make Your Own 30-Day Free Test

CHAMPION SEWER CLEANING OUTFITS

The one machine that not only cleans the sewer pipe, but in the same simple operation also lifts the dirt OUT of the sewer ON to the street. Dirt, roots, etc., are handled only once.

☛ When the bucket grabs them they come out in the bucket right to the top of the street where it empties. The advantages of this method OVER ANY OTHERS has decided the purchase of Champions by hundreds of cities and towns.

☛ The money saved on one job alone is usually twice as much as the entire cost. Almost impossible conditions are conquered with the Champion. Will clean hundreds of feet of mud, stone, branches, roots, brickbats and whatnots. Yard by yard they work their way through and haul out load after load and dump it.

Positive cleaning at minimum costs is the result every municipality wants. The Champion gets these results. It is powerful, easily transportable and swings quickly into action. No lost time "rigging up" or "digging up" with its attendant heavy expense.

Men like to work with Champion outfits because they are not subjected to the disagreeable underground work in wet, dripping manholes. With Champions all work is done from the street level.

Your cleaning crew goes about its work with no interference to traffic. Block after block is scoured clean as a whistle and your men move on to the next intersection. Really, it is almost unbelievable how much is accomplished in a day's time until one has experienced the Champion's possibilities.

THE RELIABLE BUCKET SYSTEM

Showing Outfit in Working Position in Sewer. The Expansion Bucket gets all of the deposit, regardless of the degree of congestion.

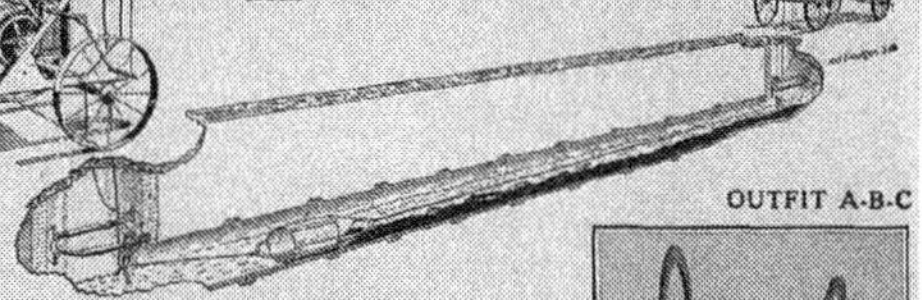

OUTFIT A-B-C

Closed Open

NEVER FAILS TO BRING UP A LOAD

Wonder Bucket Scraper

Years of experiment and experience have gone into the designing and perfecting of the Champion Expansion Bucket. It is made of rolled steel plates and is fitted with heavy, hinged steel jaws. Side plates, to which these jaws are fulcrumed, control the expansion and contraction of the bucket and, at the same time, the opening and closing of the jaws. These are sharpened that they may more easily cut into the deposit in the sewer channel. At a forward pull of the cable, they open nearly in line with the body of the bucket, which also expands. As soon as the load is taken up, the cable pull is reversed. Immediately the jaws close and the bucket contracts. The jaws are close fitting when closed and retain the load taken up, allowing only the water to escape. This feature is especially appreciated by the men on the job who, in the past, have experienced the disagreeableness of handling a wet load. To prevent excess wear, the bucket is fitted top and bottom with steel running plates.

Manhole Jack Guide for Smooth Efficiency

This device automatically guides the filled bucket out of the sewer tile and up through the manhole to the street above.

When a bucket is pulled out of the tile, it strikes the trolley arm which guides it to the center of the manhole where it passes upward and out. The trolley arm automatically locks in an upright position until the bucket is again lowered into the manhole.

The work of cleaning a sewer line proceeds rapidly and a surprising distance is covered in a single day.

Costs Less To Own Than To Be Without

Where mileage of sewers is limited the special Champion Outfit fills a distinct need. Mounted on skid brackets, this equipment is lifted onto a truck for transportation to the job and slid from manhole to manhole as the work progresses. Completely capable, its cost is so low that no community can afford NOT to have it.

Special Champion Outfit

OUR NO. D164

$389.50

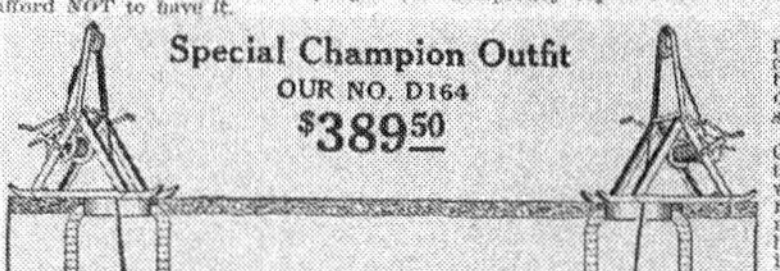

This special outfit is specially priced for small towns and includes two buckets that will take care of 8", 10", 12" or even do a good job in 15" or 18" sewers.

The outfit consists of two Champion windlasses, and a total of 700 feet of ⅜" steel cable which allows 350 feet on each reel. Two special Champion manhole guide wheels, two steel hand hooks and two steel utility hooks. One 10' pike pole with one No. 6 patented expansion bucket and one No. 8 patented expansion bucket.

No. D164. Champion Sewer Cleaning Outfit—including all of the above....... **$389.50**

You will need sewer rods, at least 350', if you haven't any please order these in addition to the above.

No. 518. 350' Champion Sewer Rods....... **$55.00**

Complete Outfit of Parts Furnished

Outfit C All the equipment listed below, which includes a full set of all size buckets for all size sewers, F.O.B. Chicago....... **$748.00**

2 Trucks with hoist windlass and swinging boom.
300 feet 3/16-in. Guide Cable.
2 Steel Cables 350 ft. long, 19 wires hemp center.
2 Kuhlman Patented Manhole Guide Jacks complete.
1 6-in. Expansion Bucket for 8-in. sewer.
1 8-in. Expansion Bucket for 10-in. and 12-in. sewer
1 10-in. Expansion Bucket for 12-in. sewer.
1 12-in. Expansion Bucket for 15-in. sewer.
1 15-in. Expansion Bucket for 18-in. sewer or larger.
1 Saw Root Cutter to fit 8-in. sewer.
2 Cable Pulleys and Hooks for gallows.
2 Swinging Boom Chains and Hooks.
2 Short Hand Hooks for manipulating buckets.
2 4-ft. Steel Utility Hooks.
1 10-ft. Manhole Pike Pole.
350 feet Special Sewer Rods, Jointed.

Outfit B All the above with 6", 8", 10" and 12" expanding buckets only. Cleans up to 15" sewers.. **$698.00**

Outfit A All the above with the 6" and 8" expanding buckets only. Cleans up to 12" sewers....... **$664.00**

Let Us Send You a Champion Sewer Cleaning Outfit on 30 Days' Free Trial

that you may judge its performance, and its value to your city, town or village, without risking a dollar. This trial will not obligate you at all. We claim our Champion Sewer Cleaning Outfit pays for itself on one job. We are never called on to prove the claim—the Outfit does that. ☛ Immediate shipments

☛ We furnish cable in any length desired. Cable in excess of the above lengths will be charged at 9½c per foot. When ordering a machine always state the size of manholes, how far apart, and also size of manhole covers. Some cities have narrow openings into manhole, and in such cases the size of the Manhole Guide Jack must be changed.

HYDRAULIC SEWER FLUSHER — Used By More Than 1,500 Cities

Clears a clogged drain or opens a sewer line in a few minutes.

The usual method is to connect the bag to the flushing hose, place it in sewer pushing it forward to the point of obstruction, and turn on the water. The heavy bag expands, conforming to the shape of the sewer and making a watertight seal. The full hydrant pressure is shot from the nozzle against the obstructing matter. In a few minutes the force of the water bursts through the accumulation and washes it down the sewer.

Order No.	Sewer Size	Hose Used	Price Each
17*	6"	2½"	$ 25.76
20*	8"	2½"	37.64
22	9"	2½"	40.00
24*	10"	2½"	46.80
25	12"	2 "	48 00
26*	12"	2½"	50.00
27	14"	2½"	65.56
29	16"	2½"	75.00
30	18"	2½"	83.20
31	20"	2½"	88 00
33	24"	2½"	117.28

*These four sizes comprise a combination very frequently ordered by municipalities.

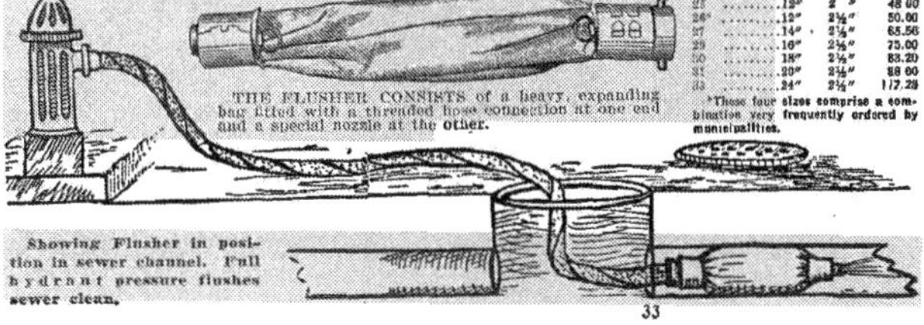

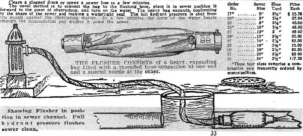

THE FLUSHER CONSISTS of a heavy, expanding bag fitted with a threaded hose connection at one end and a special nozzle at the other.

Showing Flusher in position in sewer channel. Full hydrant pressure flushes sewer clean.

The Backwater Flush

Many towns employ the backwater flush, finding this method both convenient and satisfactory. The backwater flush is made by placing the instrument in the sewer, against the flow, and turning on the water. The heavy bag expands, closing the drain and the water rises to whatever height desired in the manhole. The hydrant is then turned off and the instrument releases itself. It is withdrawn from the sewer and the whole force of the impounded water in the manhole sweeps through the channel as the flood waters above a dam sweep down a river when the dam goes out. Not only is the obstruction washed away but the entire sewer line for a distance of several hundred feet and more is washed clean.

LIGHTER WEIGHTS for smaller sewers, drains and pipes. Same Flexible Steel, UNBREAKABLE heat treated. Easy to use and serviceable for years.

No.	Size. Inches	Length 50 ft.	Length 75 ft.	Length 100 ft.
C-202.	⅛ x ¾	$2.30	$3.50	$4.75
C-203.	⅛ x 1¼	3.95	5.70	7.70

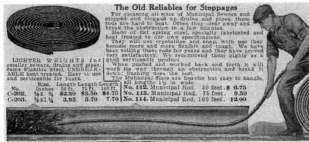

The Old Reliables for Stoppages

For cleaning all sizes of Municipal Sewers and stopped and clogged up drains and pipes, these rods are hard to beat. Often they clear away and break the obstruction in a few minutes.

Made of flat spring steel, specially fabricated and heat treated to our own specifications.

They will not crystallize and snap, with use they become more and more flexible and tough. We have been selling these rods for years and they have proved very satisfactory. We recommend them highly as a good serviceable product.

When pushed and worked back and forth it will work its way through an obstruction and break it down; flushing does the rest.

The Municipal Sizes are heavier but easy to handle, all lengths 1½ in. wide.

No. 112. Municipal Rod, 50 feet . $ 6.75
No. 113. Municipal Rod, 75 feet . 9.50
No. 114. Municipal Rod, 100 feet . 12.00

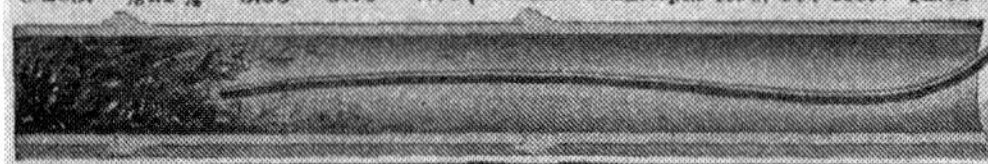

Revolving Sewer Cleaner

SPEAR POINTS

Pays for Itself First Time Used

Nothing can equal this simple, low priced device, for cleaning clogged up drain and sewer pipes.

Just bolt it to a sewer rod and it gets the rags, cloths, papers, string, etc., no matter how tightly packed or clogged. The Sewer Dept of Chicago after trying a set ordered 50 more. It surely does the work quicker and better than any other tool you ever saw.

The head is torpedo shaped and revolves on a shaft which can be readily fastened to rods. When the rod is pulled back the paper, string and other solid matter are caught by the hooks and are withdrawn with the cleaner.

No. C199. For ¾ and 1¼ in. Wood or Metal Sewer Rods. Each.......... $3.00
No. C200. For 1½ in. Wood or Metal Sewer Rods. Each........... $3.50

Spear Point Sewer Rod with Automatic Grip Handle

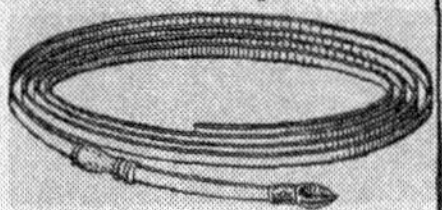

Made of a special oil tempered, flat, spring, steel wire. Will coil as shown and will straighten out and lay flat in sewer. Spear point enables user to remove stoppages easily. We also equip these rods with an automatic grip handle which enables the user to force the rods in and out of sewers and drains with ease.

	Size. Inches	L'gth 50 ft.	L'gth 75 ft.	L'gth 100 ft.
No. D637	⅛ x ⅝	6.30	7.50	8.75
No. D638	⅛ x 1¼	7.95	9.70	11.70
No. D639	⅛ x 1½	10.75	13.50	16.00

NET PRICES LESS 3% FOR CASH WITH ORDER

SPECIAL LONG HANDLE SEWER SPOONS

Made of the finest alloy steel, forged and heat treated. Formed on special pattern for the purpose of dipping mud and dirt out of the bottom of manholes and basins. With handle of second growth, air seasoned, Northern white ash, fitted tightly and completely into the blade. Heavy straps electrically seam welded to blade. Standard size, with handle 8 feet long.

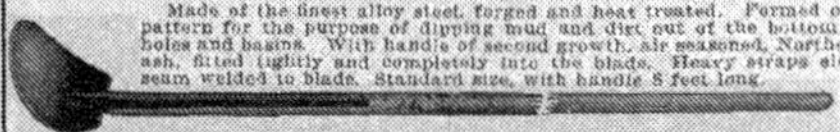

$4.95 Each

Order No. E381.
Less 10% for 12.

SPIRAL SEWER ROD HEAD

Has only one working part. Acts like an auger when pushed into the sewer pipe. The head is made of malleable iron. Spindle is hand forged.

No. C201

For ⅝" rods........$1.50
For 1¼" and 1½" rods 1.95

Flexible Steel Waste Pipe Cleaner

Handy for cleaning small clogged up service pipes and waste pipes. Consists of ¼ inch oil tempered fish tape of great flexibility which is placed in holder and arranged so that any length desired can be removed without the danger of all the tape unwinding. Holder made of flat steel bands, heavily galvanized, and acts as handle with which to grip and work the tape.

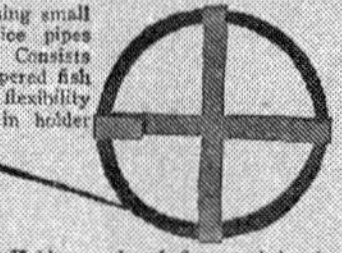

No. B974. Waste Pipe Cleaner, 25 feet long.....$1.60
No. B975. 50 feet long........................ 2.80
No. B976. 75 feet long........................ 4.30
No. B977. 100 feet long....................... 5.30

3% for Cash With Orders

Sewer Cleaning Dipper

No handle. Threaded for ¾-in. iron pipe handle. Heavy, galvanized.
No. 349. Sewer Dipper. $2.59

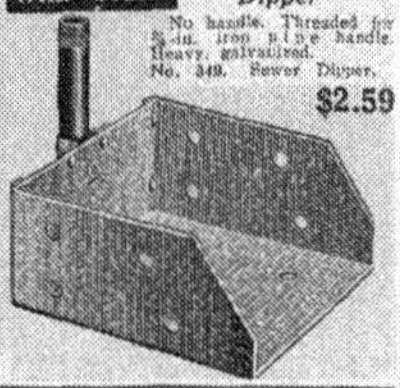

Plastico Sewer Pipe Joint Compound

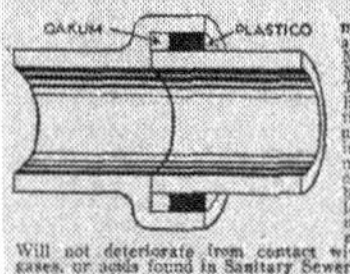

The easiest joint to make—spread on with a trowel or by hand. NO HEATING — NO MIXING — NO EXTRA TOOLS or EQUIPMENT. Water tight, stops roots, non-freezing, prevents infiltration. An all mineral compound containing no vegetable matter, made of long fiber asbestos cement mixed with mineral asphalt binder.

Will not deteriorate from contact with soil acids, sewer gases, or acids found in Sanitary Sewers. PLASTICO joints can be made in wet trenches.

No. E382. Plastico Sewer Pipe Joint Compound.

Quantity	1Gal.	2Gals.	5Gals.	20Gals.	30Gals.
Price, per gal.....	$1.00	.95	.90	.80	.80

Your Money GOES 20% TO 40% FARTHER WITH W. S. DARLEY & CO.

NEW! Darley's Snake Cleaner for Opening Sewers and Drains

2¼ in. Wide

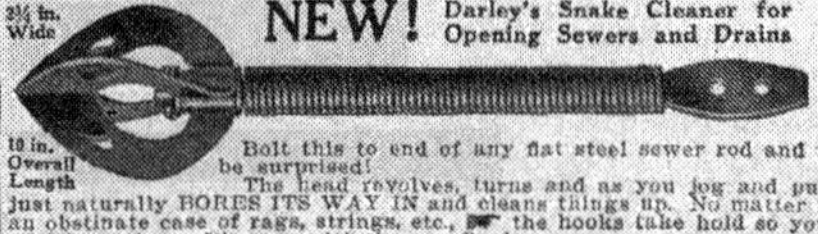

10 in. Overall Length

Bolt this to end of any flat steel sewer rod and you'll be surprised!

The head revolves, turns and as you jog and push, it just naturally BORES ITS WAY IN and cleans things up. No matter if it's an obstinate case of rags, strings, etc., ☛ the hooks take hold so you can withdraw them. It's easy with the new Snake Cleaner.

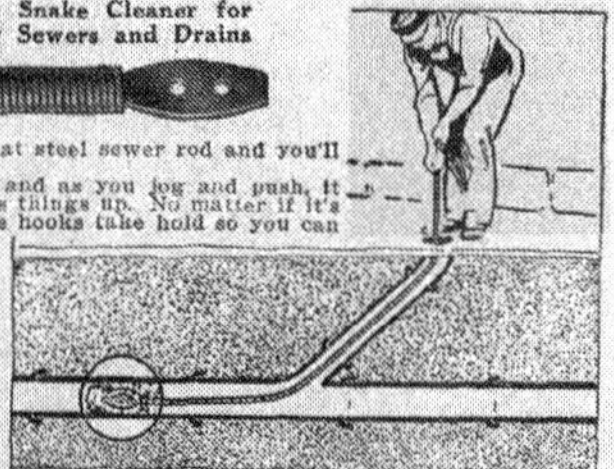

☛ It will go around a 90° angle in a 4-inch or 6-inch sewer or drain.

☛ The Revolving Steel Head is electrically welded to the spiral FLEXIBLE steel coil spring; it is very sturdy and will give years of service; long after the low price has been forgotten.

Every sewer department needs a Snake Cleaner. It will save a lot of hard work and time. ☛ We will take it back after you try it, if you don't think so. The price is so low it usually saves its cost on the first job.

No. C929. New Snake Cleaner.........$5.70

CULVERT CLEANER OUTFIT, the Only Tool of the Kind!

Strong Collapsible Scraper
Often Pays for itself
☛ First Time Used

Handle in Sections

This simple practical tool will clean any size culvert pipe in jig time. A dirt filled culvert is a problem, but not with this. One man can easily and quickly clean any size pipe from 8 inch up to 36 inch or larger. No hard labor poking or scraping away at accumulated sediment. ☛ It FOLDS as it is pushed in and comes out with a load like a HOE. The disc is cut curved to conform to the circle of pipe, so it is efficient.

☛ The handle is made up of 3 sections of our Champion Oregon Fir sewer rod. Very light, but strong and sturdy. The sections hook together and will not become unhooked in the culvert. Each section is 4 feet long, so you can reach 15 or 16 feet, or with extra rods it will reach any distance up to 100 feet.

Every maintenance crew should be provided with at least one outfit. With the handle in sections it is easy to carry and stow away and the price is so very inexpensive from Darley & Co.

No. C822. Culvert Cleaner Outfit with 3 Section Handle, only.......................... $4.85
No. C823. Extra 4-foot Rods, each.................... 65c

NEW REVOLVING ROOT CUTTER

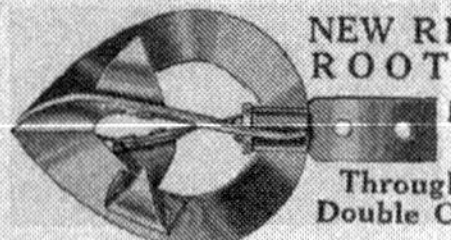

Bores and Cuts Its Way Through Roots with Double Cutting Action

Just bolt it to any size sewer rod, push it into the sewer that has the roots growing into it and then pull it out. ☛ It will bore and cut its way through the roots as you push it in and will also cut the roots as you pull it out. ☛ Each of the four blades has a rectangular knife sharpened on both edges for double action cutting. ☛ The head of the new root cutter is torpedo shaped and revolves on a shaft which can be readily fastened with rivets or bolts to long or short flexible sewer rod, any size from ⅝" to 1½" wide. The four blades, each carrying a double cutting knife, are air pointed at the nose or top and enable the root cutter to easily bore its way into the root obstruction. ☛ Made of hard tough steel, yet light in weight and easy to work with.

No. D925. New Double Action Root Cutter for ⅝" to 1½" rod, each, only.............. $4.95

W. S. Darley & Co. Offer Lowest Prices—30 Days Free Trial and ☛ 3% Discount for Cash or C.O.D.

OUR Champion sewer rods are made from selected Oregon Fir. The finest Fire Department ladders are made of this wood. Government tests have shown that Douglas Fir is the strongest structural wood now in existence and one of the remarkable things about it is the fact that it is 25% lighter than any timber of the same character.

Because Champion Rods are of Douglas Fir they are LIGHTER, much easier to handle, and they are so buoyant they FLOAT on the water in a sewer. This is a big advantage in running in rods because they glide on the water and do not need to be pushed along the bottom of the tile like the old style heavy rods.

Champion Rods are being used with much success throughout the United States and Canada. Many jobs can be done with these good sewer rods that might, if left undone, cost many dollars to do later on.

The malleable connections are made light and strong and are riveted with two flathead rivets, making a strong rod. The connections cannot come unhooked while in the tile because it requires they be held at nearly right angles to hook in or unhook. This is also an advantage while hooking together or unhooking, in taking them out. Order Champion Douglas Fir Rods for SERVICE and LONG LIFE and take advantage of the price. They should sell for more than Hickory rods but our price saves you 20% to 40%.

No. 510. Champion Sewer Rods. 3½ Feet Long. Each 63c

No. 511. Champion Sewer Rods. 4 Feet Long. Each 72c

One man can handle long runs easily.

Sewer Departments know the new Champion is the best sewer rod in the world because it not only is the STRONGEST, but because it also FLOATS.

"I've tried them all. Give me this one!"

Always in stock for prompt shipment.

Powerful Sewer Cleaning Brushes

This special brush is very successful in dragging sewers. They do not injure the laterals if same have been placed inside of the sewer, where scrapers, etc., cause a lot of trouble.

No chance of this brush getting caught in the sewer as the wire is flexible and will give under the strain.

Made of flat TEMPERED steel wire with steel protected ends and equipped with hooks for fastening on a rope or cable.

As a simple sewer sweeper and cleaner, it can't be beat. ☞ Customers that try them get such good results they order more.

Stock sizes for 4" to 15" sewer and made to order up to 36" sizes.

Made like this 10" to 15"

Sewer Brush Prices

No. D171.	4" Diam ..	$ 3.95
No. D172.	5" Diam ..	5.40
No. D173.	6" Diam ..	6.80
No. D174.	8" Diam ..	10.80
No. D176.	10" Diam ..	15.90
No. D176.	12" Diam ..	21.80
No. D177.	15" Diam ..	26.75

We Make Them to Order

☞ Made to order also for any size sewer up to 36" diameter. No. 949. Sewer Brush, to order PER INCH DIAMETER $1.90

3% Discount FOR CASH OR C. O. D.

Made 4" to 5"

MAKING THE TOUGHEST JOB EASY

We recommend the complete set of sewer rod accessories to all purchasers. You will find them practical tools, durable and inexpensive.

ROD GUIDE

SEWER ROD GUIDE, No. 512. Every order for sewer rods should include this item $1.10

CORK SCREW ROD Pulls out rags, paper, tree roots. No. 514 .... $1.40

GOUGE

SEWER ROD GOUGE Used to remove cement, roots and other obstructions which impede sewage flow. No. 516 .... $2.25

COLLAPSIBLE SCRAPER ☞ IT FOLDS as it is pushed in and comes out with a leaf like a HOE. The scraper should be with every rod outfit. No. 513. Small size for 6-in. and smaller sewers $2.75
No. 513B. Large size for 8-in. and larger sewers $2.75

Sewer Suits ☞ Complete Protection

Made of Olive drab double texture materials, with fine quality rubber joining the two fabrics. Body of jacket made one piece to reduce number of seams, hood with drawstrings around front, storm-fly front, closing the garment perfectly. All seams sewed, strapped and cemented, making it absolutely waterproof. Sleeves at wrist made to button close to exclude water, mud, filth.

Pants made with minimum number of seams, draw-string at waist, bottom of pants can be buttoned close to exclude sewage and for wearing inside boots.

Thoroughly waterproof, completely protected, most comfortable combination ever designed for cold wet work.

No. 583. Sewer Suits. Each $6.95

☞ Note: Regularly made in 3 sizes which correspond to 38, 44 and 48 inches breast measure respectively. Give size wanted when ordering $7.95

☞ Root Cutting Steel Sewer Saws

Every Sewer System should have this tool. Gets the roots and pulls them out. The most effective sewer tool made. Used by Sewer Depts. from coast to coast. ☞ Use this and you will soon forget that root infested sewers were ever a costly problem.

Made of steel bent to angle shape with special steel cutting blades riveted to place at 45° angles across the holder. Cutters sheared to angle points and laid criss-cross to produce a sawing effect on all sides when pulled back and forth. When roots are cut, torn and loosened they come out with the cutter.

The cutter complete, 2 units, is 6 feet long, with 12 strong blades, each 6" across. ☞ Suitable for sewers of any size from 6 inches up.

No. 473. Root Cutting Sewer Saw, weight 31 lbs., each $16.80

Boot Protectors

Lighten the work and prevent sore feet. Used in excavating. Made of heavy steel, fitted with webbing strap and will fit any boot or shoe, to which it adds life. Sherardized finish, which makes them rustproof.

No. C824, Boot Protectors, each . . 67c

For Entering Manholes Filled with Deadly Gas

Many deaths have occurred when sewer workers have been overcome with gas, which would have been prevented by this simple mask. The worker can enter a manhole or a sewer and breathe fresh air out of the nose of poisonous gas vapors.

This mask fits over the mouth, nose and eyes. Fresh, pure air is received through a tube valve; exhale air passes from mask through a valve port.

The tubing is so flexible it can be tied all up into knots and the natural air supply cannot be shut off. Practically indestructible, having a spiral wound wire on the inside and is water and air proof and stopping on same will not injure it. The tubing is carried by a belt around the waist instead of by the head of the worker and the mask can be put on and taken off as conveniently as a baseball mask, allowing the wearer to do his work without danger or discomfort and his breathing is natural. Head straps are made from flexible webbing and adjustable to any size head.

You can enter any manhole, sewer, etc. filled with gas or vapors of any kind and shut off valves or make repairs.

The complete equipment includes the mask, head harness, 25 feet of flexible tubing, plated intake funnel, also belt, one-piece airtight facepiece with goggles for the protection of the eyes, all packed in store case with handle, lock and key.

No. 274. Safety Hose Mask. Complete $37.50

Extra Tubing, any length desired, with coupling attached .... 60c per foot

NOTE—If you order more than the 25 feet of hose specify what length you want.

Yours ☞ 3% Discount for Cash or C. O. D.

Very Handy Suction Pumps

These pumps are especially suited for lifting muddy water from holes while leaks are being repaired and for quickly draining trenches, or excavations, manholes, valve boxes, etc., after rains or storms. Every Water, Gas, Electric and Sewer department should have one or more hand pumps ready when needed or called for. These are made just for that purpose. Strong, extra heavy, metal riveted, as illustrated, with single clapper valve and leather plunger. They lift a lot of water at every stroke and saves trouble or gutter pipe can be hung on the spout to carry off the discharge if desired.

	Length Measured Over All	
No. 673.	2-inch diameter by	$6.80
	2½ feet.	
No. 674.	2-inch diameter by	$7.75
	3 feet.	
No. 675.	3-inch diameter by	$8.35
	2½ feet.	
No. 676.	3-inch diameter by	$9.65
	3 feet.	

69c Manhole Cover Lifter

A handy hook lifter for raising manhole lids. Many municipal departments are using them. It's easy to lift manhole lids with this hook lifter. Made of a good grade of octagon steel, painted black. Point is ¾" wide and carefully tempered.

No. D178. Manhole Cover Lifter, ⅝ by 18", each only 69c

Clean, Sparkling, Granite-Concrete Drinking Fountains
You Save 50% At Our Low Prices

Attractive New Designs—Polished Smooth

Constructed of attractive, smooth, Granite-Concrete and designed to meet the popular demand for a durable fountain at a most reasonable price. Note the pleasing simplicity of line in the designs; the bright Chromium plated valves and bubblers add a further degree of beauty—a beauty which will be indefinitely retained.

Most artificial stone fountains have such a rough surface that in a few weeks grime and dust discolor them. As public fountains get a good many hard knocks we recommend our new polished smooth Granite-Concrete instead of enamel. Enamel, so dead white, is more suitable to bathrooms, where the surface isn't liable to get chipped off. Granite-Concrete is to be preferred—in our opinion, regardless of price—for outdoor use and the color harmonizes better with surroundings in schools, public buildings, institutions, etc. Order Granite-Concrete and you can be quite sure of approval. Our Granite-Concrete composition is made of black and white granite concrete, water polished by special process to produce a beautiful terrazzo finish.

Granite-Concrete foundations will not chip or break and can be left in service throughout the year. Thoroughly reinforced with steel bars. Holes are provided in base for bolting to foundation.

$48.50

The Washington

This is especially suitable for parks, stadiums and golf courses, etc. It is much lower in cost, more attractive than any similar design we know of. Base is square, 28 inches in diameter. Column is star shape, 12 inches in diameter above base, tapering to 11 inches below top. Top is square, 36 inches in diameter. Supply and return pipes cast directly in concrete. We furnish this complete and 4 angle stream SANITARY bubblers and piping as illustrated. The shipping weight of the Washington is 1,000 lbs.

No. 635. Washington Fountain with angle stream SANITARY Bubblers, each........**$48.50**

☞ With Self-Closing Valves under the angle stream SANITARY Bubblers, Only, each........**$68.10**

The Dixie   The Lincoln

$14.95   **$21.35**

Suitable for street corners, highways, parks, etc. Supply and return pipes are cast directly in concrete. Both the Dixie and the Lincoln have very pleasing designs and we furnish them complete with angle stream SANITARY bubblers, either continuous flow or self-closing piping, so that either can be installed in an hour by any good mechanic.

No. 632. The Dixie. Weighs 150 lbs.
With angle stream SANITARY Bubbler....**$14.95**
With Self-Closing Valve under the angle stream SANITARY Bubbler.... **19.60**
No. 633M. The Lincoln. Weighs 233 lbs. **$21.35**
With angle stream SANITARY Bubbler.
With Self-Closing Valve on side and angle stream SANITARY Bubbler.... **24.85**

MODERN SANITARY BUBBLERS

Our new Granite-Concrete drinking fountains are fitted with the newest type SANITARY bubblers, heavily chromium plated. The finest fixtures made.

Especially designed to meet the specifications of the Sanitary codes in all States. Two types, continuous flow or with built-in self-closing valve.

The new bubblers throw an angle stream. Bubblers are located to prevent back suction of the water from the bowl. Deep bowls prevent splash and overflow.

BUBBLERS SOLD SEPARATELY

For all makes of fountains. Chromed. Made with ⅜″ I.P. male or ⅝″ I.P. female threads.
No. D218. Sanitary Angle Bubblers, each........................**$3.65**

W. S. Darley & Co. Offer Lowest Prices—30 Days Free Trial and ☞ 3% Discount for Cash or C. O. D.

Hydrant Attachment Makes Street Shower for Hot Days

Refreshment, health and recreation for the children! Ever watch 'em, the shouts of laughter, the little screams of joy from the tiny tots, as these suffering hot days, when kindly authorities allow a fire plug to be turned on, just for a little while!

How much better the fun—study it. Not a full hydrant 2½″ flow but a SHOWER that takes so little water you will never miss it. The volume of the spray can be regulated, to cover semi-circular area up to 50 or 60 feet in diameter—a gentle shower of cool, clear water, so well diffused that even very cold water is tempered by the warm air before it falls on the little bodies of the children.

Low enough in price so that ordering a few for the children won't strain the budget of any City, Town or Village. ☞ We'll help—if there is no ready money, and take a warrant in payment.

No. D745. ☞ Made of high quality brass castings, carefully machined and nickeled. Aluminum coupling nut. Weighs only 4 lbs. Diameter of head 5″. Overall length 6″. Spray volume regulator included. Can be put on or taken off in 20 seconds.

Remember the hot days when you were a kid and how you would have enjoyed this fun!

Furnished with straight connection to hydrant, or with the swiveled connection for hydrants that have outlets to one side. Say which you want. ☞ Ask the Fire Chief for size of hydrant opening and the kind of threads on it. ☞ Or he will know. ☞ Or the Chief can give to fit your HYDRANT THREADS. The Chief will know. ☞ Or the Chief can give you sample fire hose coupling to send us.
No. D745. Street Shower, straight tailpiece, each....................**$9.65**
☞ In lots of 5 or more, $8.55.
No. D746. Street Shower. Swiveled connection can be adjusted to any angle, each.......................................**$10.65**
☞ In lots of 5 or more, $9.55.

New Playground & Wading Pool Showers

No. D747 Playground Shower has semi-circular head like the Street Shower shown above. Tapped for 1¼-inch pipe. Can be bushed for any size pipe.
No. D748 Wading Pool Shower has full circular head than head. For connection to a 1-inch vertical pipe standard. May be bushed for any size pipe.
No. D747. Each............**$4.85**
No. D748. Each............**4.25**
In lots of 5 or more, $4.35.

THE NEW COOPER CLIPPER POWER LAWN MOWER

A Great Value At Our Special Low Price

$84.50

The new "Clipper" is all that you expect a fine, full-powered mower to be. It is a high grade, full size power cutting unit, which is a model of simplicity and is priced low to make it the greatest value in a fine power mower.

It is a complete power mower using the new Briggs and Stratton motor and a heavy duty Worcester Quiet-Mo (Special) cutting unit. A sturdy, quiet running, light weight power mower for fast, efficient grass cutting.

Motor operates the reel and propels the machine over the lawn. Easily started with rope starter. ☞ Or can be supplied with foot starter. The fully adjustable, 13″ ball bearing reel makes short work of the average acreage. Best quality rubber tires. Handles easily.

"Hinged Power" eliminates the necessity of clutches and a very simple take-up for reel, chain and belt, reduces all wearing parts and service adjustments to the minimum.

Our Complete Line of Power Lawn Mowers Includes a Model for Every Type of Service. Write Us.

Cast aluminum alloy deck, on which the motor is mounted, prevents deflection and distortion of the entire mowing unit. The Cooper "Clipper" is light in weight and easy to handle. Instant and simple adjustment to vary speed according to individual requirements of the operator is provided.

The "Clipper" has been thoroughly tested under all possible mowing conditions, and has proven without a question, its claim for finest engineering design, stamina, and quiet, efficient operation. Its price is much lower than you would expect to pay for a power mower of such high quality.

No. D710. New Cooper Clipper Power Mower, complete, at our special low price, only **$84.50**
☞ Can be supplied with foot starter for $3.50 extra.

3% SAVINGS FOR CASH WITH ORDERS OR C. O. D. SHIPMENTS

39

Powerful Hand Torches *For* Economical Weed Burning

FOR BURNING WEEDS along roads and streets, in vacant lots, fields, around Fence Posts, Wire Poles, and along Fences, Barns, Warehouses, Railroad Tracks, etc. ALSO IDEAL FOR CHARRING FENCE POSTS.

FOR BACKFIRING to control forest fires. Used by U. S. Forest Service, National Park Service, U. S. Department of Agriculture, State Foresters, etc.

FOR BURNING WEEDS AND GRASS in and along Driveways, Park Walks, Tennis Courts, Cemetery Walks, Golf Courses, etc.

FOR BURNING THE WEEDS in and along Irrigation Ditches and Drainage Canals to facilitate the free flow of water and prevent weed seeds from spreading.

Complete Ready for Work Including Tank, Pump, Gauge, Valve, Burner, Hose, Carrying Strap
All for Only $14.75

★ OUR VALUE LEADER ★ NEW WEED BURNER ★ ONLY $14.75

An unbeatable value in a complete weed burner, sturdily constructed for long hard service. Construction and design the very best. A marvel in simplicity and a perfect wonder in operation.

Produces a powerful, windproof, smokeless flame 2″ in diameter and fully 30″ long. Terrific temperature of 2000 degrees. Anyone can operate it. The powerful heat is always under control. Consumes only ½ gal. oil per hour.

Economical all-around efficiency. The most efficient all purpose torch value ever offered.

For BURNING WEEDS and Weed Seeds in flower and truck gardens; between rows of flowers, plants, trees and vegetables; in lawns, hedges, rock gardens and seed beds; in driveways, garden walks, stone walks between flagstones; on tennis courts, playgrounds, athletic fields; on golf greens, tees and fairways; any kind of weed . . . any place . . . any time.

For HEATING Water, Tar, and Asphalt. For THAWING Frozen Pipes, Pumps, Switches, Outdoor Machinery and Equipment, Frozen Ground, Watering Troughs, etc.

For MELTING Ice and Snow on Platforms, Roofs, Pumps, Ladders, Scales, Trucks, Culverts, Manholes, Concrete Forms; Coal, Sand, or Gravel Piles.

For DESTROYING Dead or Diseased Fowl or Animals; Grasshoppers, Locusts, Mosquitoes and Insect Pests of all Kinds; Poison Ivy, Sumac, Dogberry and other Poisonous Weeds.

AND . . . . for charring fence posts and poles, preheating truck and tractor engines in cold weather for easier starting (the 2000° heat PRECEDING the flame does the trick); preheating for welding and brazing; melting lead and babbit; shrinking and expanding metal; heating pipes up to 2″ diameter for bending or straightening . . . . and many other practical, year-round uses.

Burns kerosene, gasoline, range oil, stove or light furnace oil. Includes four-gallon fuel and air tank, tested at 100 lbs., made of extra heavy copper bearing galvanized steel, with all seams and fittings welded (not soldered); equipped with a large 2″ diameter brass air pump, which has brass (not rubber) check valve, pressure gauge, snap on shoulder strap, 3 feet high test special oil resisting rubber hose with brass hose nipples and clamps and a new seamless steel coil burner and windshield with bronze regulating valve. Extra drop handle.

complete, only $14.75

For Thawing and Heating

ONE YEAR GUARANTEE: Every Burner is unconditionally guaranteed for the Whole Year. Any part showing a defect within one year after purchase will be replaced without charge. Guaranteed workmanship and materials.

TAKE IT ON 30-DAY FREE TRIAL. So confident of your complete satisfaction we offer this to you on trial without any obligation on your part. Any official is welcome to try one.

Lowest Prices—30 Days Free Trial and ☞ 3% Discount for Cash or C. O. D.

Heavy Duty Weed Burners

For destroying weeds, seeds and brush, spot and strip burning, mosquito and hay fever control, thawing and melting. Burners are fitted with adjustable windshield. Detachable flat flame nozzle furnished if specified.

No. D856. Consists of 3-gal. welded steel fuel tank with back rest and adjustable shoulder straps, 6-foot oil hose and unions, combination fuel valve and strainer. Burner fitted with adjustable angle handle. Flame, 3x16 in., 2000° F. Burns kerosene or light distillate. Fuel consumption, 1½ g.p.h. Net weight, 27 lbs. Shipping weight, 43 lbs. Complete, each $42.75

Same without Back Rest or Straps, each $37.75

No. D857. Consists of 5-gal. welded steel tank with side handle, 12-foot oil hose and unions, combination angle carrying handle. Flame, 4½x30 in., 2000° F. Burns kerosene or light distillate. Fuel consumption, 1¾ g.p.h.

Burner weight, 9 lbs. Shipping weight, 50 lbs. Complete, each $52.50

Handy Weed Burner With Flat Flame

For permanently destroying weeds in driveways, foot-paths and along fence-rows. A sure means of eradicating Mesquite Brush, Johnson Grass, Bind Weed, Quack Grass, Witch Grass, Chick Weed, Thistle or any kind of worthless vegetation. Burns the stalks, seeds and all.

Can also be used for burning and destroying dead game, burning rubbish and underbrush, fighting forest fires, etc.

Produces a powerful, flat blue flame 6 in. wide and 26 in. long, 2000° F. Special burner has patented flat coil which makes a flat flame—which is double effective for weed burning. Nothing equals our flat flame burner; other burners use flat nozzle attachments which restrict flame and burn off frequently. Shipped complete with 1½ gal. capacity welded steel Pressure Tank with 90 lb. gauge, 7-in. heavy brass Air Pump, bronze Filling Cap with Air Release, Oil-regulating Needle Valve and a seamless steel, guaranteed FLAT FLAME BURNER. Shipping weight, 24 lbs. Consumes about 1 gallon of kerosene per hour.

No. D855. Handy Weed Burner with Special Flat Flame, each, only $26.85

Used as a Thawing Torch for thawing out frozen water, steam and oil pipes, thawing out frozen culverts, melting ice and snow off scales, ladders platforms or any outdoor machinery.

30 Days APPROVAL AND FREE TRIAL GUARANTEES YOUR SATISFACTION

Heavy Duty Weed Burner Outfit on Wheels

15-gal. welded steel fuel tank mounted on detachable hand truck; 28 feet oil resisting hose with unions, combination fuel valve and strainer. Burner has extension handle and adjustable shoulder strap. Flame, 4½x30 in., 2000° F. Burns kerosene or light distillate. Fuel consumption, 1½ g.p.h.

Burner weight, 13 lbs. Shipping weight, complete outfit, 140 lbs.

No. D858. Heavy Duty Weed Burner Outfit On Wheels, complete $89.25

Same outfit, not mounted on wheeled truck $78.75

Powerful New Weed Burner Hand Torch

A powerful flame for destroying weeds, seeds, brush, germs, garbage; for preheating for welding, heating for shrink fits. Burns kerosene, coal or furnace oil. Weighs only 18½ lbs. Burner only, weight, 3 lbs. Burner fitted with combustion tube; starts in 5 min. Burns in any weather with a steady, clean, reddish blue flame 3 in. by 30 in. long, 2000° F. Consumes only 1½ gal. oil per hour. Snap shoulder strap for easy carrying.

Complete with 4-gal. WELDED steel tank, powerful quick action pump, pressure gauge, full funnel filler, tank handle, 7-ft. oil resisting hose with brass nipples, fully enclosed seamless steel coil burner and windshield, fitted with bronze fuel regulating valve. Wonderful for thawing frozen pipes and ground in winter. A general utility torch useful for many purposes.

No. D847. Powerful Weed Burner Hand Torch, only $17.75

Only $17.75 Complete

FLAME 3x30 in. Long

BURNING WEEDS IS MORE ECONOMICAL THAN CUTTING

When you cut weeds you only remove the stalks, stems and leaves—next year's crop of weeds will be more dense because you have done nothing to eliminate the seeds on the ground nor the roots in the ground. When you burn weeds, you not only remove everything above the ground, but also the seeds and roots on and under the ground, thus preventing future growth and giving you real, lasting weed control. Do not try to burn weeds after a rain. It is more economical to do your burning after a few days of dry weather.

BEST TIME TO BURN WEEDS is in the spring, while the weeds are still dead or dormant, for then you can destroy not only the young weed sprouts but last year's crop of old weeds before they come to life again. Or Burn Weeds in the Fall, when the weeds are nearly wilted and dying, thus making it easy to burn them. Besides by burning in the Fall you destroy the Weed Seeds, thus making sure that new weeds will not sprout again the following year. You must destroy the seeds on the ground for real Weed Control.

We Are Always Ready to Give You Service ★ Telegraph Day or Night at Our Expense

40

Beautifully Chased Designs By Skilled Craftsmen

Your Choice
$2\underline{15}$ NICKEL PLATED

Your Choice $2\underline{70}$ NICKEL PLATED | SAVE MORE Send Cash Take Off **3%**

Chrome Plated 25c Extra	Any lettering Any State Seal You Specify
Gold Plated $1.00 Extra	All Badges Have Safety Clasp Fasteners
Rolled Gold $5.50 Extra	

THE ILLUSTRATIONS ARE ONE HALF THE ACTUAL SIZE OF THE BADGES

WARDROBE AND UTILITY CABINETS
Quality Steel Construction at Low Prices

Our cabinets offer maximum value in Style, Convenience, and Durability—whether for use in office, home, factory, school or wherever space saving storage facilities are needed. All have nickeled hinges and door handles; smooth rolled finish edges. They're finished in brown or green enamel, to harmonize with any interior. All steel, electrically welded, with three point locking device and padlock attachment. Also with flat key lock, and in walnut grain finish at low extra cost.

DOUBLE DOOR WARDROBE CABINET
Plenty of room for several sets of clothing and general storage of equipment. Coat hanger and hat shelf. 72 in. high, 20 in. deep and 64 in. wide. Coat compartment 60 in. high, with 9 in. shelf at top. Shipped Knocked Down only.
No. E353. Double Door Wardrobe Cabinet.....$17.90

STORAGE CABINET
Equipped with four shelves for storage of miscellaneous equipment. Shelves spaced as follows: Top 15 in., second 9 in., all others 12 in. Height 63 in., width 22 in., depth 12 in. Shipped Set Up.
No. E354. Storage Cabinet.................$10.95

SINGLE DOOR WARDROBE CABINET
This single door cabinet is a popular model. It's equipped with coat hanger and hat shelf. 72 in. high, 20 in. deep and 22 in. wide. Coat compartment 63 in. high, with 9 in. shelf at top. Shipped Knocked Down only.
No. E355. Wardrobe Cabinet...............$13.55
Any cabinet furnished with flat key lock and 2 keys for $1.10 extra. Special cabinets built to order. Send us size and shelf arrangement for quotation.

No. E353.   No. E354.   No. E355.

Oak Swivel Office Chair
$8.95

Made in plain oak with quartered oak top slat. Finished in light oak, dark oak, fumed oak, and brown. Width of seat, 20 in, depth of seat, 17¼ in., height of back from seat, 18½ in. Seat, back posts, arm posts and base made out of 1½ in. stock. Arms are made out of 1¼ in. stock. Rail, top and middle slats, 1 in. stock. Pressed steel iron, guaranteed 20 years. Casters at no extra charge.
No. C93A. Oak Swivel Office Chair, each, only $8.95

$3.60
Holds 1000 Letters

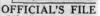

Steel Office Desk

A big, good-looking, substantial desk for the busy official, office worker or superintendent. A desk that helps a man get out his work and keep track of things. Three drawers at your right hand, three at your left hand, and a wide center drawer which locks.

A place for everything, which means order and better work. This desk has a steel top and is finished in the same beautiful neat olive green as the cabinets shown in this catalog. The appearance of a bank president's all-steel efficiency desk at fraction of the cost; that's what you get in our Utility Desk.

No. E356. Steel Utility Desk, with Top 28x58 in., as shown. Shipping weight, 220 lbs.......... $27.90
No. E357. Steel Utility Desk, with Top with drawer on one side only and middle drawer, with Top 28x43½ in. Shipping weight, 190 lbs....................$18.90

TELEGRAPH W. S. DARLEY & CO.
ANY TIME ABOUT ANYTHING

OFFICIAL'S FILE

Cold rolled steel case, baked enamel dark green crinkle finish. Chrome locks, handle and label holder. Printed folders and A-Z index. Or you can label folders as you wish.
Safe compartment for confidential documents built into top with lock different from the file lock.
Expanding front for easy filing. 11¾ in. high, 12¾ in. wide 6¾ in. deep.
No. E391. Only........... $3.60
No. E397. Similar to the above file but without compartment in the top. Strongly reinforced red rope fibre. A-Z expanding index.
Only......................$3.89

Special NEW Catalogs
Sent FREE

This catalog, chock full of wonderful values, while representative of our entire line of equipment, covers only in a general way what we have to offer and is by no means complete.

Hence if your present needs are not shown herein, request the Special Sales Book covering your particular needs—it will be mailed to you immediately.

That's why we publish these special catalogs—to give you all the information you want before you buy.

52 PAGE CATALOG OF
POLICE SUPPLIES

84 PAGE CATALOG OF
FIRE EQUIPMENT

45

Extra fine STEEL LETTER FILES *At New Lower Prices*

THIS FILE Only $14.25 — Made of steel, the logical material. No warping, sticking, cracking. Highly resistant to fire. Wonderful strength and rigidity insured by welding all members of the frame. **THIS FILE Only $20.90**

Finished in soft-toned olive green with handles and label holders of solid brass, harmonize perfectly with any other furniture.

Drawers operate on a progressive roller-bearing side-suspension—opening and closing smoothly, easily, silently, even when heavily loaded. They cannot be withdrawn accidentally; two safety catches must first be released.

The adjustable follow-blocks are smooth and positive and hold papers always in position.

4-Drawer Unit. The standard size in most offices 14¾ inches wide, 52 inches high 25 inches deep. No. D444. 4-Drawer File, Letter size. Complete, only **$14.25**	**Combination Unit.** The two top drawers are fitted with removable partitions and can be used for 3x5 or 4x6 cards. The 3 lower drawers are letter size. No. D446. Combination Unit. Complete **$20.90**
No. D445. 4-Drawer File, Legal size $18.95	
For lock controlling all drawers add $5.55 to above prices.	With lock controlling all drawers $26.90

Above Prices are for GREEN   For Walnut or Mahogany add $5.55 to Prices

Document File
4 Drawers Each with Three Divisions
$37.75

Walnut or mahogany $4.08 extra. Lock $3.75 extra. The same high grade quality as our standard letter files. 14¾" wide, 52" high, 26" deep. Each drawer has 3 compartments. No. E393. Olive Green, only **$37.75**

Card Files
Only $1.50
Finest furniture steel, smoothly finished. Rubber feet. Drawers slide easily. Spring compressors. Combination handle and label holder of brass. Olive green baked enamel. Filing space for 1000 cards in each drawer. May be stacked together.

One Drawer Cabinets		Two Drawer Cabinets	
For 3x5 Cards No. D447 ..... $1.50		No. D454 ..... $2.75	
For 4x6 Cards No. D448 ..... 1.95		No. D455 ..... 3.95	
For 5x8 Cards No. D449 ..... 2.50		No. D456 ..... 4.95	

Locks $1.00 per drawer extra. Brass, 10" high, $2.95.

A Money Saving Value!
$11.89   No. 1 Quality Cabinet
Roomy—66x14x18"

Constructed entirely of cold-rolled steel by skilled craftsmen. Extra strength and durability are welded into this quality grade storage cabinet.

Five adjustable shelves.

A handsome piece of office furniture. Smoothly finished all around. Green enamel finish is baked on inside and out—the equal to an expensive automobile finish.

No. E394. A $17.50 Value, special low price only **$11.89**

Vest Pocket Adding Machine
Adds up to a billion (999,999,999), is guaranteed trouble proof and is simple enough for a child to operate. Ideal for adding bills, statements, purchases, for checking records, and reports, and for use in homes, offices, schools and factories. Constructed entirely of steel and brass. Size 8x3½x3¼.

No. E360. Vest Pocket Adding Machine, only **$2.45**

CORNERS AND JOINTS SMOOTHLY FINISHED

STURDY SHELVING

JIFFY SNAP BRACKETS

PIN TUMBLER LATCH AND LOCK

CENTER SHELF FIXED FOR RIGIDITY

SHELVES ADJUSTABLE HALF INCH SPACING

REINFORCED DOORS

Only $11.89

Double Door Cabinets

Cat. No.	Size In.	Wgt. Lbs.	Price		Cat. No.	Size In.	Wgt. Lbs.	Price
D462	66x18x78	180	$27.25		D474	30x18x78	171	$24.90
D463	36x24x78	226	30.75		D475	36x24x78	198	28.70
D464	36x18x78	164	24.25		D476	24x18x78	139	22.98
D465	24x18x78	152	23.75		D477	24x24x78	157	24.57
D466	24x24x78	174	25.48		D478	24x18x66	122	20.60
D467	36x12x66	140	24.25					
D468	24x15x66	123	20.33					
D469	36x18x66	134	20.95					

Roomy Steel Storage Cabinets. Four shelves. / Wardrobe Cabinets. Hat shelf, full width coat hanger rod, two rust proof coat hooks, double doors. Clean protected storage for the garments of several people.

3% Discount for Cash or C.O.D. Orders

STEEL LOCKERS $5.90
For Municipal Offices, Fire and Police Stations, Community Buildings, Hospitals, Jails, Institutions, Factories. No. D483. Lockers, green.

Wid.	Dep.		Ht.		Price Per Locker
12 x 12 x	60 inches	$5.90
12 x 15 x	60 inches	6.24
12 x 15 x	60 inches	6.96
15 x 15 x	72 inches	7.25
15 x 18 x	72 inches	9.55

All above prices for lockers knocked down, for lockers ship-ped set-up add 10%. Prices are per locker in groups of 3 or more wide. Add 5% if only two lockers are ordered. Add 10% if only one locker is ordered. Above prices include padlock attachment, add 45c per locker extra for Yale built-in lock and 2 keys.

We can supply double tier lockers, box lockers, closed base lockers, 2 person, 4 person, 6 person lockers, basket racks, key cabinets for swimming pools and gyms. Write us.

WRITE FOR QUANTITY PRICES

Combination Padlocks
A general purpose padlock. Scientifically tested, resisting picking, moisture, and tampering. Combination. Three number with approximately 216,000 possible changes. No. D191. Padlocks, each **92c**

You Save 50%!
Triple Lockers
Only $11.80 for 3

A sensational value made possible only by large quantity production during slack seasons. We don't know how long we can continue to fill orders at this price—better order now.

Strong and durable ALL STEEL construction. Overall dimensions, 34" wide, 12" deep, 61" high. Each locker unit 11" wide, 12" deep, 61" high giving ample room for coats, hats, boots, tools, etc. Each locker compartment has hat shelf, single point locking device, padlock attachment, 3 coat hooks.

Sanitary base, no place to catch dust and dirt.

Finished in olive green. Shipped knocked down only. Easy to erect.

No. E395. Triple Locker Unit, 5 lockers for the price of one, our special **$11.80**

WRITE US FOR LOWEST PRICES ON SPECIAL SHELVING AND BINS

48

BEST OFFICE EQUIPMENT AT LOWEST PRICES BY W. S. DARLEY & CO.

Steel Office Cabinet
With 14 Drawers

$10

☞ Each drawer 4⅛" deep. Entire cabinet all steel construction, finished in dark olive green lacquer.

For Conveniently Holding

Letters, Correspondence, Catalogues, Stationery, Cancelled Checks, Office Materials,
Reports, Folders, Clippings, Pamphlets, Instruments, Stock of Parts, etc.

Stands 38½" high, 21" wide and 14½" from front to back.

No. D180. Cabinet **$10.00**

Seal
This seal you can hold in your hand and use it like you would a ticket punch. The "NEW ALUMINUM" is made very strong—yet, weighs only ten ounces. It is the ideal seal to use when away from your desk as it can be carried in the pocket.
A leatherette pocket carrying case is supplied. **$4.75**

No. B684. Seal and Case ....
PRICE INCLUDES ALMOST ANY PLAIN LETTERING. Add $3 for State Coat-of-Arms.

Now You Can Have a Private Telephone System ☞ at Very Low Cost

Here is, without a doubt, one of the most sensational new items ever offered Municipalities, Fire Departments, Water Works, etc. A private telephone system, that costs practically nothing to operate, that will talk and ring for thousands of feet. Until you actually have your own private telephone system, you cannot fully appreciate the time and delay it will save to give orders or obtain information.

Desk Type Telephone

Never before have complete, modern, private telephone systems been offered at anything approaching these low prices. They have a standard transmitter and receiver and are about two thirds the size of big telephones.

Wall or desk type telephones are furnished as desired, the desk type being furnished with a concealed buzzer, and the wall type with a concealed bell. Depressing the plunger on which the handpiece rests operates the signal of the other telephone.

Two wires are required between the telephones, and two dry cells at each telephone. Each telephone is furnished complete with terminal strip, 50 feet of single conductor interior wire, and full instructions for installing.

Full instructions for wiring are supplied with each system. Anyone with a screwdriver, a hammer and some staples can install these systems in a few minutes time. No soldered connections are necessary. No electrical knowledge is required. Merely follow the simple instructions and the system is ready for service.

These systems are for private service only. They are not designed for connection to public exchange systems.

Prices shown are for single telephones. For the 2 phone system it is necessary to order 2 phones at $6.00 each. For the 3 phone system it is necessary to order 3 phones, one of which must be a desk phone.

$6.00 *Wall Type Telephone*

2 Phone System
No. D181. Desk Telephone, ea. **$6.00**
No. D182. Wall Telephone, ea. **6.00**
☞ Above prices include 50 feet interior wire.
No. D183. Extra inside wire, per 100-foot roll **60c**
No. D184. Outside wire, per 100 feet **90c**

3 Phone System
No. D185. Desk Telephone, ea. **$2.50**
No. D186. Wall Telephone, ea. **7.00**
☞ Four strand wire required for this system at these prices.
No. D187. Inside wire, 50 ft. **$1.25**
No. D199. Outside wire **3.75**

LETTER FILE Constructed Entirely of Steel

Amazing!

Only $11.90

Strong—Neat —Serviceable

For Office and Transfer Use

Lacquer Finished In Dark Olive Green

Drawers roll easily on steel rollers

Adjustable follower furnished with each drawer

Case 40½"x15¾" x25½".

Drawer 24½" x 12¼"x10".

SPECIAL
No. D179. Letter File with four drawers **$11.90**

Manila Alphabetical indexes—20 division 46c per drawer EXTRA

3% Discount FOR CASH OR C. O. D.

Alphabet GUIDES
25 Divisions ☞ A-Z
No. B509.
Letter Size 85c
No. B510.
Legal Size.....1.10

Shipping wt. 128 lbs.

Super Values AT LOWEST PRICES AND 30 DAY TRIAL WITH 3% DISCOUNT For Cash

Manila Folders
No. B511. Letter Size, per 100 **$1.90**
No. B512. Legal Size, per 100. **$2.50**

Register Boxes for Patrolmen

This system has no complicated wiring, and yet it will register the call at the box and give the exact hour, day or night that the box was "pulled."

The Register Box contains a 24-hour clock movement, with a 24-hour paper dial. This part of the box is locked up with a Yale key kept in the possession of the Mayor or Chief, or Clerk. The party having possession of this key opens the box once a day and collects the records and puts in a new one.

The other section of the box, which is a separate compartment, is opened by the Patrolmen's key. The Officer opens it and inserts his Register Key and pulls the box. The time is registered on the paper dial. The Officer carries nothing but his keys.

The keys of each Patrolman are different. You can install one Call Box, or seventy-five. You can have one pair of keys, or you can have seventy-five pairs of keys. Each pair of keys will be different, but each one will work any box to register the Officer that pulled it.

You can start by ordering one box, or two boxes, and one set of keys for one Night Watchman and then you can add as many more as you wish, any time. The keys are practically impossible to duplicate and indestructible.

POLICE REGISTERS

No. 816. Register Boxes, complete with 1 year supply of 24-hour charts. Each ... **$85.00**
No. 817. Patrolman Keys. Per set **$2.00**
No. 818. Extra charts. Per box of 1 year's supply **$2.75**

The No. 816 Box includes a winding key and an opening key, but not a Patrolman's key. For Patrolman's registering key specify No. 817, which includes a registering key and an opening key.

Police Patrol Register

A police registering system with "Call Stations" or Key Boxes, for the Night Watch in Towns and Villages. Your homes and places of business must be watched to be safe, not only for theft but for FIRE.

☞ Equip your town with Key Stations, in the residence and business districts and police or watchmen will most efficiently perform their rounds and duties and give their best services to the protection of life and property from dark till dawn.

The clock contains a paper dial on which is recorded the time, when the station key is inserted. The chart or dial records are kept by the Chief, or Marshal, or Town Clerk, and can be referred to at any time.

The clock is admirable! Its cast aluminum case is very light, though exceedingly strong. It is further protected by a stout, genuine sole leather waterproof pouch, and slings over the watchman's shoulder by a comfortable strap.

The distinctive grille guards the dial of the clock yet permits easy reading with the hands in any position. The durability of the "Alert" clock is almost beyond belief. Many of these clocks are in service today which have been in constant use since 1908.

One clock will register 33 different key stations. This system can be started with just a few stations and more added at any time. Let us send you a clock and as many stations as you want, for a month's trial. The best proof will be the satisfaction it gives on the job.

KEY STATIONS

No. C670. Patrol Alert Clocks.
Each **$50.00**
No. C671. Key Station with Box.
Each **1.75**

48

Telephone Fire Alarm System
Sirens Operated From Any Distance

When the telephone operator gets a "FIRE" call she simply plugs in the connection to the fire dept. phone. Say it's No. 13; she knows it.

Instantly the phone box at the fire dept. rings and a little bit of the current closes the CONTACTOR, which closes the Control Switch that starts the Siren.

☞ It is so simple it is a wonder somebody didn't think of it before!

But it took W. S. Darley & Co. engineers to design the system so it works perfectly. It is so sure, so dependable that when installed you can forget it. It will need no attention, oil or adjustments for years.

No changes to make in the telephone; when the siren starts the first fireman to reach the phone just lifts the receiver and asks the operator where the fire has started.

Regardless of how you operate your siren now or what devices you have in use this system can be connected independently WITHOUT DISTURBING THE OTHER.

No. 13520. Automatic Telephone Fire Alarm System consisting of new model independent type Magnetic Control Siren switch and our Double Action Automatic Telephone Contactor switch for all sirens up to 7½ H.P......... **$49.75**

No. 13521. The Same only for sirens of any size over 7½ H.P...... **54.75**

In order to supply you we must be given the make and Horse Power of Siren, the voltage and if 3 phase or single phase current is used and the cycles. See name plate on Siren or ask your local electrician.

And ask the Telephone Exchange what is the Ohms resistance and the bell ringing voltage, or energy, on the Fire Station telephone and if the telephone circuit is A.C. or D.C. ☞ Also have them SPECIFY THE RESISTANCE of the telephone bells on the same line as fire station phone or residence phone.

☞ Our engineers must have all the above information in order to meet your requirements.

Your Siren Is Started Automatically When the Operator Rings the Fire Alarm Phone

WIRES to Siren

REMOTE CONTROL SWITCH

W.S. DARLEY & CO.

WIRES to Control Switch

WIRES to Contactor

PHONE BOX IN ☞ FIRE DEPT.

← No. 13-13

CHAMPION LITTLE GIANT FIRE SIREN

$98 75
Complete Ready to Install

Enclosed Weatherproof

The sheet steel housing protects the Little Giant completely and is designed to modern airflow principles. The lower louvres in the housing are made for easy flow of incoming air, allowing unobstructed airflow. The upper louvres are of opposite type to allow easy flow of outgoing sound waves, which go out unobstructed in all directions. Behind the louvres is fine mesh copper cloth. Nothing to obstruct air or sound but practically sealed-safe from rain, snow, sleet, hail.

Special bracket for hoisting the siren to the roof or tower.

Weatherproofed in building, our motors will stand any exposure and will even start entirely submerged in water. Fully enclosed in the housing and doubly protected by its own weatherproof construction.

Universal type motor for any 110 volt AC or DC current. No line transformer needed.

Built with super HIGH SPEED special siren motors, so powerful that they turn faster by hundreds of revolutions per minute. The tune is HIGH-PITCHED and remarkable for distance and penetration.

Vertical Type, Ball Bearings and 2 Years Lubrication Sealed in.

Rated as ¾ H. P.
But Actually
Develops Over
One Horse Power

Champion aluminum rotors turn friction-free on ball-bearings. And they are dynamically balanced.

The Little Giant is built to the same high standards as the larger 2½ and 5 H. P. Champion Sirens, with the same 5 year guarantee.

An ideal fire siren for small towns, villages, mills, camps, industrial plants, who wish to have the advantage of siren protection but who do not need a big siren.

Can be installed and wired in a few minutes. Can be heard up to half a mile, but many say the distance is often far greater.

No. E750. Champion Little Giant Fire Siren, ¾ Horse Power. Our special low price................ **$98.75**

Make Your Own 60 Day Showdown Test at Our Expense

3% for Cash With Orders

Factory Howler

Vibrating type, ample volume, penetrating tone, distinctive pitch.
No. E404. For 110 volts A.C... **$9.90**

CODE CALL SIGNALING SYSTEM

Operates mechanically. Calls 20 different people—gives call 1, 2 or 3 times on one setting. Just set the dial at the number to be signaled, push the plunger, and the call is given automatically. For use with siren, bell, horn, buzzer or light. For factories, mines, camps.
No. E402. Code Caller...................... **$39.90**

CODING SIREN

Our No. C978 Siren, shown above, made special for use with No. E402 Code Caller. Powerful code blasts cut through worst noise.
No. E403. Coding Siren, for 110 volts........ **$59.50**

SET THE DIAL ... PUSH THE PLUNGER

Powerful Compact Fire Alarm Siren $37 50
For ☞ 110 Volts and only

For small Villages and Communities this compact electric siren will prove a most efficient fire alarm.

Operates from any 110-volt line socket; just plug it in.

Being absolutely weatherproof it can be placed outdoors and one or more can be mounted on roofs or any convenient location. Any 110-volt button or switch will get it going. You have to hear this siren to appreciate it. The speed is about 7,000 RPM. Very fast, clear and high pitched tone.

Many towns use this as an advance alarm siren to clear the way along certain congested routes for fire apparatus. It's good.

The high siren tone cuts through all other noises and it may be adjusted to direct the sound to the best advantage.

The special Universal motor operates on A.C. or D.C. The extra powerful aluminum rotors and the motor will give years of service with little or no attention whatsoever.

Finished in a brilliant fire department red.
No. C978. Fire Alarm Siren for 110 Volts. Each... **$37.50**

☞ Even the smallest villages should not be without some kind of fire siren protection. It won't cost you ten cents a month for current. Later on, if you wish, this can be traded in for a large size Champion—we will take it and make you a good liberal allowance. It does not seem possible that anything for so little ($37.50) could be loud enough, but just try it on suspicion; no harm done and no hard feelings if you return it to us.

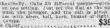

Automatic and Remote Control Combined With Automatic Time Stop

All fire sirens are 100% better and more efficient if AUTOMATICALLY CONTROLLED by this new device. It will operate any siren, of any make, from 1 H. P. to 12 H. P. At the touch of a button (which may be any distance away) it first starts the siren and keeps it automatically running UP and DOWN the scale for 3 MINUTES, and then stops it.

The Unit, mounted in a steel wall cabinet, is an assembly of a relay, starting train, interrupter, relay and an ELECTRIC MOTOR, a marvel of compactness and inventive skill. The cost of operation is nothing, perhaps only 1 cent for 100 sirens. We can't imagine it will ever break down or need any attention.

Your present or any remote control switch, to which this is connected, will do. The same push or station buttons can also be used; or we will supply these items at small cost.

No. D258. Electric Auto-Siren Control **$34.50**
In steel cabinet.

☞ Made 6-second ON and OFF contacts or any timing desired. ☞ Test your siren and see how many seconds it takes to reach the high siren tone; then set the switch and let it coast for exactly the same number of seconds, then call us and we will time it accordingly.

It Pays to Buy Your Siren From Darley

There's no secret about the great popularity of Darley's Champion Sirens among wise, careful buying Fire Chiefs and Municipal Officials. They know Darley will not take chances of disappointing customers and they know Darley's method of Selling and Shipping Direct-to-Thousands-of-Customers is the Most Economical Way. That's why W. S. Darley & Co. is the World's Largest Business of its kind ... selling Direct-by-Mail ... and Saving You 25% to 40% on quality Sirens.

49

Neither Snow • Nor Rain • Nor Cold • Stays Champions

5 Horse Power $272⁵⁰

Including Remote Control and Start-Stop Station

CHAMPION ENCLOSED

Engineered and Built By W. S. DARLEY & CO.
The Established Leader of All Siren Makers

New and improved enclosed design, with practical weatherproof construction; not the only enclosed sirens BUT THE BEST. Undoubtedly, the finest, most powerful Sirens made in America, at an amazing price. We are the pace-makers for all siren builders. Year after year W. S. Darley & Co. sell more sirens than the total output of several makers put together.

Compare prices and you will find that other makers want to get from $360.00 to $490.00 for sirens of equal horsepower. The difference is in the SELLING PRICE and NOT in quality, workmanship or horsepower.

If we did business in the usual way and priced our 5 H.P. Siren at $410.00 we would be reserving for agents, salesmen or dealers, a commission of 33⅓% or $136.67, making the billing price to them $273.33. **We have no salesmen, agents or dealers to protect and will bill you this siren direct at the amazingly low price of $272.50.**

Every Bit As Good As They Look

The new Champion Enclosed Weatherproof Fire Sirens are every bit as good as they look. Their enclosed construction and glistening red finish are all anyone sees but in your own mind the many features built into Champion Sirens are even more important. Features DESIGNED into them from the drawing board up, from the laboratory out. Qualities that will show in faithful service during the siren's long life.

Scientific design and construction make the Champion weatherproof. The fully enclosed sheet steel housing protects the Champion completely and yet is designed to modern airflow principles. The lower louvres in the housing are made for easy flow of incoming air, allowing unobstructed airflow. The upper louvres are of opposite type to allow easy flow of outgoing sound waves, which go out unobstructed in all directions. Behind the louvres is fine mesh copper cloth. Nothing to obstruct air or sound but our new Champions are practically sealed-safe from rain, snow, sleet, hail.

Special bracket for hoisting the sirens to the roof or tower.

The motor is doubly protected. Weatherproofed in building, our motors will stand any exposure and will even start entirely submerged in water. Fully enclosed in the housing and doubly protected by its own weatherproof construction.

Buy Your Siren for Radius Range

Many siren purchasers are fooled on the difference between distance range and radius range. A siren may have an advertised distance range of 1 mile, or 2 or 3 miles. What this actually means in plain English is: That that siren was heard 1, 2, or 3 miles in one direction —the direction the wind was blowing when the test was made. Such "distance" tests are usually made in fair weather.

Radius 2 to 5 Miles

But you know and we know that fire is no respecter of fair weather nor the direction the wind is blowing. It strikes anywhere in any weather.

So we build Champions not for distance range but for radius range. Which means our Sirens are designed to THROW OUT blanketing sound volume to all parts of the compass—as spokes shooting out in all directions from the hub to the rim of a mighty wheel. To the north, south, east and west —everywhere within the sound radius its warning cry is heard. To every home its protecting alarm brings confidence.

NO ONE, not even salesmen and engineers who survey your towns, can foretell with any degree of certainty, the performance of any particular siren in your surroundings. That is why we ask you to accept a cheerful 60 Day Champion Demonstration at our expense. By making your own tests you will KNOW the performance of a siren in your town. You won't have to speculate or take anyone's word for it—you will KNOW the siren's performance against such factors as the surrounding buildings, topography of your locality, type of installation, location, etc.

We are the pioneers of this method of selecting a siren and we have satisfied thousands of Officials, intelligent and conscientious men like yourselves, who purchased Champions because they wanted the best protection and the greatest value for their town.

Underwriters Approved Fire Sirens

From the Faithful Performance of Their Duty

WEATHERPROOF FIRE SIRENS

Wonderful Super-Speed Motors

Men who know, thrill with enthusiasm to the motors we build into Champion Sirens. Motors with abundant surplus power for their job. Motors capable of starting in a split second! Motors which accelerate from nothing to 3,600 revolutions per minute in three seconds! Unbelievable, almost. But a "Champion FACT" nevertheless, made real by the perfect co-ordination of every Champion part with every other Champion part.

2½ Horse Power **$169⁵⁰**

Including Remote Control and Start-Stop Station

The Motors are built in our own shops. Unlike other siren makers we do not equip our Champions with stock motors by various makers, that do very well for operating pumps, churns, blowers. Other siren makers may use good motors (good for the purpose intended)—but not equal to our own make Champion motors. The Ball Bearing Shaft is not new because this principle has always been employed by us, but shafts are larger and bigger and unbreakable.

The rotors in Champion Sirens are one piece aluminum castings with extra large sound ports that give a piercing shriek in all directions. Their speed is 3600 RPM, with capacity for thousands of cubic feet of air per minute.

Vibration is so practically and entirely eliminated that in shop tests at full speed of 3600 RPM these NEW Champions are not even bolted down—"they stand without being hitched."

Champion's great aluminum rotors turn so friction-free on their ball-bearing mountings that they revolve at the touch of a child's finger. They are machined to micrometrical standards of 1/10000 of an inch. And they are dynamically balanced!

Champions are so utterly free of friction-drag that lubrication would seem almost unnecessary. Yet, exceptional provision is made for it. A three-year supply of lubricant is sealed into every Champion Siren!

Scientific design eliminates as many as EIGHTY PARTS in Champion Sirens—parts which encumber other sirens—parts which run up costs and run down performance.

Superior design, finest machining through every step of manufacture, absolute precision standards, finest steels and aluminum—these are the actual-fact reasons for the phenomenal performance and everlasting life of every Champion Siren.

More Costly to Build

TRUE, it is more costly to build the Champion way—more costly by far than general siren practice. Dynamic-balancing alone in Champions calls for many hours of skilled highest-paid hand work on each siren. And machining moving parts to less than a hair's breadth clearance is not done in a flash.

But CHAMPION PRICES—horse-power for horse-power, sound-volume for sound-volume—are lower than any others and much lower than most others.

For we buy the finest steel and aluminum for Champions in advantageous quantity purchases. And we design, pattern, engineer and manufacture Champions, from motor cores to gleaming enameled exteriors, within our one organization, in the biggest and best equipped siren factory in America.

Because Champions are the finest and most powerful sirens made, they have been the choice of hundreds of Fire Departments.

Champions have such extraordinary motors and great sound ports and power, they have been listed as STANDARD by the Underwriters' Laboratories and APPROVED by the National Board of Fire Underwriters for Fire Department service in all parts of the United States.

Champions out-perform every other make. Champions start in a split-second—accelerate to full speed in three seconds—outdistance all others with their mighty roar—and last practically forever.

When the First Call is sounded it instantly transmits to your Firemen, that a fire has started—NOT when the whole building is a mass of flames but at the start of the fire.

Seconds count when a fire starts; one flame inhaled means death. Every minute unchecked it destroys property. No Champion ever delays one second in starting and its sound travels with the speed of radio.

With lightning speed its distinguished warning cry overcomes the barriers of fog, heavy atmosphere, wind, rain, snow, sleet.

Not only in brilliant performance but in value as well, the Champions set new standards. No other siren maker is in a position to give the value W. S. Darley & Co. give their customers.

DOUBLE 5 YEAR GUARANTEE

We are in business for the protection of our customers' interests. And we guarantee every Champion Siren for 5 years and after that to replace FREE any defective parts needed.

Radius 1 to 2 Miles

We Ask You to Accept a Cheerful 60-Day CHAMPION Demonstration at Our Expense

Write or telegraph us, at our expense, to ship your city, town, village or community a Champion Siren on approval and sixty days' free trial.

Give your citizens Champion protection for sixty days—use and test the siren in every way. Compare it with any other siren made, regardless of price. Judge its beauty, distinctive tone, tone volume, radius range. If the Champion does not sell itself you will not be asked to buy; return it at our expense without correspondence or conversation.

2½ HORSE POWER Champion Fire Department Siren, for 110 or 220 Volts A.C. SINGLE PHASE only, including Remote Control Switch and one Station Push Button. No. E400...................................$169.50

5 HORSE POWER Champion Fire Department Siren, for 3-phase, 60-cycle, 220 Volts, including Remote Control Switch and one Station Push Button. No. E401.....................................$272.50

No. E401X. This is the same as the above except with SINGLE PHASE motor........$318.75 Blue Prints showing easy wiring diagrams for your local electrician, are included.

☞ Extra Station Push Buttons, you can use as many as you want and start and stop the siren $2.50 from several locations, each...................

See next page for Coding Controls and Alarm Boxes.

Easy Payments if you Wish. We can offer you 2, 3, 4, 5 or 6 months' time to pay. ☞Or a YEAR if you will divide it into EASY PAYMENTS.

When a Fire Call Comes in the Operator Just Touches a Button and the Champion Roars for Men and Apparatus

at America's Lowest Prices

51

Darley Saves You Over 50% On

DARLEY'S RITEFLEX CHICAGO

RITEFLEX

A quality hose with finest rubber tube and heavy, long wear cotton jackets. The jackets are circular woven with the filler strands running squarely around the tube and completely covered by the warp strands which are woven in a close weave. We use only selected grade extra long staple cotton in the jackets. Our loom weaves both jackets at once so that both are perfectly uniform and resist all tension together. By using the finest cotton yarn and weaving it more efficiently we are able to produce a more flexible and stronger hose but with less bulk and weight.

FLEXIBLE—PLIABLE—COMPACT

Riteflex Hose rolls up easily and you can pack 50% more of this hose in a body than any other hose on the market. Its flexible and pliable and easy to handle. Because the jackets are woven together Riteflex resists higher pressures. It has a burst pressure of 1200 lbs. When under pressure it lies straight and does not twist and elongate.

Riteflex stands 600 lbs. test pressure and will give several years of dependable service under severe conditions.

It's a super quality fire hose produced with the first consideration that it be flexible, roll up compactly and easily, pack in a body using less space. Made of finest cotton and pure Para rubber by modern methods, Riteflex is equal to the most expensive hose for the years of faithful service it will give.

No. E541. Riteflex Fire Hose, 2½ inch, single jacket, per foot..........**68c**

No. E542. Riteflex Fire Hose, 2½ inch, double jacket, per foot..........**78c**

No. E541P. Riteflex Fire Hose, 2½ inch, WITH PROTECTOSEAL TREATMENT, single jacket, per foot..........**75c**

No. E542P. Riteflex Fire Hose, 2½ inch, WITH PROTECTOSEAL TREATMENT, double jacket, per foot..........**85c**

No. E546. Riteflex Fire Hose, 1½ inch, single jacket, per foot..........45c
Double jacket, per foot..........55c

No. E546P. Riteflex Fire Hose, 1½ inch, WITH PROTECTOSEAL TREATMENT, single jacket, per foot..........45c
Double jacket, per foot..........57c

PROTECTOSEAL TREATMENT

Protects hose against over mildewing, rotting, freezing, in our process the compound is applied under scientific control and saturates every strand of both jackets. Does not affect tube. Moisture is effectively sealed out—Protectoseal hose sheds water like a duck's back. The jackets cannot absorb water and become water soaked.

After fires our hose does not have to be dried out but simply dried off. The best hose for Volunteer Depts., which usually don't have facilities for drying hose and can't afford to have a reserve hose supply.

Laboratories have made searching tests on Protectoseal hose. After being soaked in water and exposed to moist, humid, warm air for over six months, Protectoseal hose showed no mildew formation, nor discoloration, nor loss of strength. Protectoseal treatment is a permanent protective quality of Darly hose. It will not wash out.

Protectoseal hose sheds water and cannot freeze even in subzero fire fighting.

Both Cable Cord and Riteflex hose are made with Protectoseal treatment.

THE FINEST HOSE

Long years of experience have taught us what Fire Chiefs like and don't like in fire hose. Our mill with this knowledge and experience has developed the new Cable Cord and Riteflex Fire Hose. Two brands that Chiefs like to use; hose that stands up under hard service; hose that lasts year after year under abusive conditions.

Our mill is weaving the jackets for Cable Cord and Riteflex hose on special looms, which weave both jackets at once. This makes both jackets absolutely uniform, which is impossible when the jackets are woven separately and one jacket pulled thru into the other. Both jackets are woven in perfect conformity with each other and fit closely and smoothly with even tension at every point, resisting all pressure strains uniformly by expanding together. Double looms permit weaving the jackets in opposite directions so that the pressure tensions exerted in opposite directions on each jacket are balanced.

BALANCED WOVEN JACKETS

Balanced weaving results in greater pressure resistance and prevents buckling, twisting, writhing and excessive elongation or stretch. Our loom automatically maintains a uniform tension on all the cotton it weaves into the jackets. This is very important because unless uniform tension is maintained the hose will not stand high pressure.

Balanced woven, Cable Cord and Riteflex hose will lay straight where other hose lays snaky. The result is: Easier to handle. Being straight, water passes through with less resistance and less friction loss—it conserves pressure and gives a better fire stream. Each jacket is circular woven—the filler strands (those running around the hose) being completely covered and protected by the warp strands (those running lengthwise). In weaving the jackets we use 25% more warp strands than other hose makers. Cable Cord and Riteflex jackets have greater pressure resistance and stand more abrasion and dragging in actual service.

Our superior weaving produces lighter, more flexible and stronger jackets. Super-strength jackets, with less bulk and less weight but with greater strength and burst resistance.

Only the finest grade cotton goes into Darley fire hose. Unlike most hose makers, we do our own twisting. All the cotton is twisted uniformly and with equal tension, resulting in stronger strands and in turn stronger jackets with greater burst resistance.

The yarn used in the jackets exceeds the Draper standard for fire hose yarn by 50% in tensile strength. Our yarn is made from extra long staple cotton which gives it much greater tensile strength and enables us to make hose having very high bursting pressures and great durability.

SLOW AGING RUBBER TUBE

The tube, of course, is the most important part of the hose. The tube we are speaking of and the best grades of fire hose, never actually wears out but is subject to deterioration. A fire hose tube may be made of first grade materials, under scientific control, and to exacting standards in order to withstand aging and deterioration.

The tube in our fire hose is made of the finest Para rubber, of 6 carefully calendered sheets. Each sheet is minutely examined to detect the slightest defect. After inspection, the sheets are plied together and vulcanized into a solid, homogeneous whole. Small pin holes are infrequently found, yet never to be entirely eliminated in rubber sheets of such great length. Should there be a small pin hole in any one ply the other plys seal it during vulcanization. Therefore no defects can cause the tube to break down or wear out. Our method of making 6 ply tubes, neatly but jointed, eliminates air bubbles, pin holes and imperfections that cause blowouts under pressure. Beware of seams.

FREE SAMPLES — Cut from regular stock. GENUINE. Nothing fancy and not "made-up"; just a real honest-to-goodness sample to show you the through and through QUALITY of our hose.

Super Quality Fire Hose

THAT MONEY CAN BUY

less tube fire hose, which some makers use because it is much cheaper to manufacture. But in making the seamless tube there is no way to detect imperfections except under very high pressure and often imperfections do not show up until the hose is in actual service.

Only new selected grade Plantation rubber is used in the tubes. The raw rubber is carefully washed and dried before being compounded with the chemical ingredients. The rubber of itself has no commercial value as it is too plastic and soft and lacks mechanical strength. It must be compounded with expensive chemicals in the exact proportions computed by our chemists. All ingredients must be carefully weighed after which they are put in a metal container for mixing and compounding. Tests are carefully made of each batch to be certain that the compounded product is accurate to specifications.

Our chemists and engineers not only experiment, test and make recipes, they supervise the manufacturing and inspection.

We do not use any rubber substitutes, fillers, or reclaimed rubber in our tubes. Tubes so made are usually undercured to give the appearance of a fine rubber stock. But nothing can equal our tubes made of pure, all Para, live, resilient, slow-aging rubber. Beware of undercured hose, it's made of inferior grade rubber compound. It will deteriorate much quicker, giving several years less service.

NEW IMPROVED VULCANIZING

Our chemists and engineers are most exacting in the vulcanization and curing process of the tube. In vulcanizing, the tubes in the jackets each length of hose is cured on an individual steaming head to eliminate subsequent distortion when under hydrostatic pressure. In this operation the tubes and jackets are stretched under the steam pressure and the inspectors can instantly detect any defect. The back of our rubber tube completely fills the valleys between the cotton strands. This insures a smooth, non-corrugating tube under pressure. Friction loss is held to a minimum. Adhesion between the tube and jacket far exceeds all specifications.

Vulcanization is the combining of sulphur with rubber and other ingredients under heat. Vulcanization is carried out at high temperature and the cure is obtained by steam under pressure. Vulcanizing must be done quickly in order to reduce as far as possible the deteriorating effect of heat upon the combination of rubber with sulphur. Our chemists have developed organic accelerators which greatly reduce the time of vulcanization and improve the quality and life of the tube.

LONG LIFE TUBE

Darley chemists have solved the problem of lengthening the life of the rubber tube. Vulcanization, as carried on formerly, created a tendency for the rubber tube to harden and lose its elasticity. Our chemists discovered antioxidants which greatly prevent deterioration of the tube. Although costly, liberal amounts of antioxidant are used to insure long life and years of service for our hose. The antioxidants protect the tube for years against natural aging with no loss in tensile strength. Our tube has a tensile strength of over 2400 lbs. Because it is properly compounded with the correct proportions of sulphur, rubber and antioxidants, all vulcanized scientifically, our tube will retain its useful life for many years. The tubes in Cable Cord and Riteflex hose age five to ten times slower than other tubes. They will stand the stresses and strains when hose is loaded and unloaded and carried folded in bodies. Our laboratory is constantly testing and checking our tube to improve it. By all known accelerated aging tests our fire hose tube is far ahead of any other.

CABLE CORD

In making the new Cable Cord Basket Weave Fire Hose, the first requirement is strength and durability under hard service. Only the best grade extra long staple cotton goes into the jackets—only selected grade Para rubber is used in the tube. Nothing is skimped or saved. A super service hose to meet the severest requirements.

Cable Cord hose is made not merely to stand up under hard service but to give years of faithful service under abusive conditions.

A hose for the hardest service in the larger cities and for the Chiefs in smaller towns who want the best hose at a money saving price. A hose which will outlast the most expensive hose by years.

In our modern mill we weave the inner and outer jackets of Cable Cord hose on one loom at the same time. By weaving both jackets in one operation we are able to produce absolutely uniform jackets which resist all tension strains together. The jackets stand higher pressure and have higher burst resistance because they are woven uniformly. Cable Cord hose has a burst pressure of 1200 lbs.

Cable Cord hose is made with the filler strands running squarely around the tube and with the warp strands running lengthwise in basket weave. The warp strands in the jacket are spun and cabled, making the strongest cords it is possible to produce. Cable Cord hose is the strongest, most rugged and most durable hose on the market.

The cabled cords or strands are like the steel cables used in bridge construction. They are made on the same principle to obtain unusual strength and wear resistance. Each warp strand in Cable Cord hose is made up of four separate strands all cabled into one stout compact strand. Each of the four strands is made of four threads of selected grade, long staple, cotton yarn, twisted on our machines.

The yarn in Cable Cord hose is twisted three times as much as the yarn in single straight twisted strands. This extra twisting makes the yarn compact, firm, solid, stronger and more resistant to wear and abrasive dragging.

Tests conducted by the laboratories and Fire Department Officials show that cabled cord jackets will stand the abrasive wear of dragging three times as long as straight twisted single strand jackets.

The cabled warp strands take all the abrasive wear of dragging and fully protect the filler strands. The fillers are straight twisted to stand the pressure strain.

Cable Cord Hose stands 600 lbs. test pressure.

For the longest wear, ability to take abuse, sheer endurance, the Chief or Official buying fire hose for his town, will want to consider Cable Cord Hose carefully.

No. E443. Cable Cord Fire Hose, 2½ inch, single jacket, per foot.................... **72c**

No. E444. Cable Cord Fire Hose, 2½ inch, double jacket, per foot.................... **82c**

No. E443P. Cable Cord Fire Hose, 2½ inch, WITH PROTECTO-SEAL TREATMENT, single jacket, per foot..... **79c**

No. E444P. Cable Cord Fire Hose, 2½ inch, WITH PROTECTOSEAL TREATMENT, double jacket, per foot.................... **89c**

No. E445. Cable Cord Fire Hose, 1½ inch, single jacket, per foot............................53c
Double jacket, per foot............................55c

No. E445P. Cable Cord Fire Hose, 1½ inch, WITH PROTECTOSEAL TREATMENT, single jacket, per foot 47c
Double jacket, per foot............................50c

COUPLINGS We will attach your old couplings FREE. If you want NEW COUPLINGS we will give you genuine UNDERWRITERS Pin Lug Couplings for Double Jacket hose at $3.90 per set and for Single Jacket at $3.20 per set.
Or we can give you Underwriter Rocker Lug Couplings for Double Jacket hose at $5.20 per set and for Single Jacket at $4.54 per set.
Specify thread or send sample coupling.

RITEFLEX and CABLE CORD HOSE MADE FLAT

For compact folding and loading. Because fire hose is in the flat position 98% of the time it should collapse easily without strains and tensions that affect the life of the rubber tube. We cure our tube flat so that the hose assumes natural flat position when not under pressure. In service the hose takes a round shape for maximum flow and pressure.

53

You Can Handle 99% Of Your Fires With 1½ Inch Leader Line Hose!

Thousands of Chiefs use 1½" hose and know its advantages. But there are thousands of Chiefs who do not have any 1½" hose in service in their Depts. And it is these men who concern us because we know the great advantages of 1½" hose and we like to tell our friends about a good thing. We were one of the pioneers of the 1½" hose, recommending it to our Fire Chief friends over 10 years ago. We introduced it to thousands of Depts. and we're still spreading the good word.

If you don't have 1½" hose, the biggest improvement you can make in your Dept. is to order this hose from us at once. Many Chiefs have written us enthusiastic letters saying their new 1½" hose actually doubled the efficiency of their Depts. We have shipped hundreds of thousands of feet of 1½" hose on our established terms of 30 Days' Free Trial and Approval but no Chief ever returned a single length of hose to us. Every Chief who trys 1½" hose can't see how his Dept. ever got along without it.

The 1½" hose is so light and easy to handle that your men will surprise you the way they can lay lines and get water on the fire with speed and efficiency.

One man can easily handle a 1½" line under pressure, moving it along without great effort. A 1½" line is speedily taken up a ladder or up stairs.

You can fight 99% of your fires more effectively by using 1½" hose lines in place of the heavy, hard to handle 2½" lines. Two or three men with 1½" hose lines can move around easily and quickly, knocking out fire and keeping it from spreading. Using 1½" lines you can confine a fire by surrounding it and prevent it from spreading; you can get to the seat of the fire without losing precious seconds when your men ventilate. For a fire that threatens to become serious a 2½" line can be used on the heart of the blaze and two or three 1½" lines used to attack the fire from different angles, to confine it and prevent it from spreading.

Many Chiefs underestimate the effectiveness of 1½" hose lines, regarding them like booster lines. Other Chiefs are successfully using 1½" lines with ⅜, ⅝, or ¾ tips more and more in place of heavy 2½" lines. Scientific fire fighting requires more use of 1½" lines which can be moved around quickly, can be shut off if not needed and will cut down the damage caused by using too much water. The modern Chief realizes his duty is to save property from all damage and accordingly he aims to use no more water than he needs to extinguish fire.

We diagram at left some of the effective layouts of 1½" hose lines which are used by Fire Chiefs all over the country. You can double your efficiency by using 1½" hose and save money on first cost and depreciation. For the biggest improvement in your fire fighting order some 1½" hose from Darley.

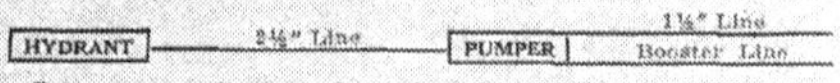

Drive pumper direct to fire and go into action with booster line. Lay 2½" line to hydrant. Lay 1½" line from pumper to fire to back up booster line. The 2½" feeder line from hydrant supplies adequate water for this layout.

Drive pumper direct to fire. Pull off 1½" hose and 2½" hose in siamese layout. Drive pumper to hydrant, laying out 2½" line. Connect pumper to hydrant with suction hose. (Using our No. E399 Siamese in this layout three 1½" lines can be taken from the 2½" line.)

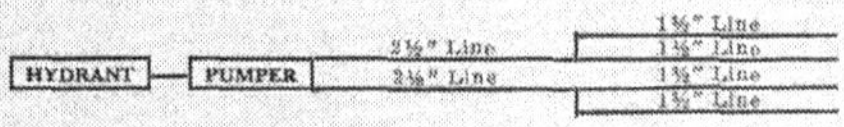

In this layout the pumper takes the hydrant in the usual way and two 2½" lines are laid from the pumper to the fire. Two 1½" lines are siamesed from each 2½" line.

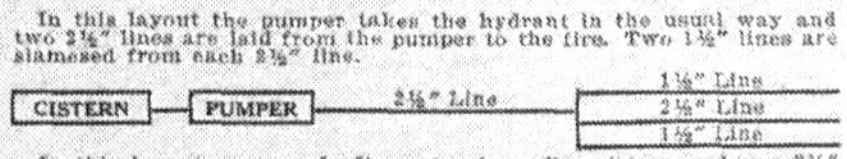

In this layout pumper drafts water from fire cistern and one 2½" line is laid from pumper to fire. With our No. E178 Siamese two 1½" lines and one 2½" line are taken from the single 2½" line.

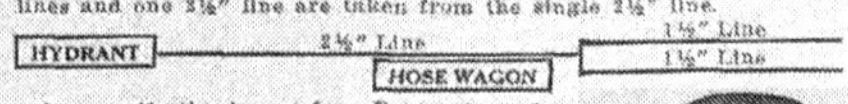

A very effective layout for a Dept. using a hose wagon not equipped with pump. Or for a town with good hydrant pressure this layout saves the time required to hook up a pumper to the hydrant.

COMPLETE OUTFITS OF 1½ INCH HOSE WITH TWO NOZZLES AND SIAMESE

Regular 2½" Fire Hose→ From Pumper to Fire

Included With Every Outfit! THE FINEST LEADER LINE SHUTOFF SIAMESE See Page 12

You Get Two Colt Shutoff Nozzles With Rubber Bumper Tips With Your Darley Leader Line Outfit

Complete Leader Line Outfits

Consisting of Siamese, Nozzles, Hose, Couplings and Gaskets

No. C660. With 4 lengths, 200 feet, of Champion Special 1½" Leader Line Hose, single jacket, with Rocker Lug Couplings attached; 2 Colt Shut-off Nozzles, 1 Special Shut-off Leader Line Siamese and Gaskets for all connections.......... **$109.00**
With double jacket hose.......... $139.00

No. C661. With Underwriters Leader Line Hose, single jacket.......... **$129.00**
With double jacket hose.......... $147.00

No. E447. With Riteflex Leader Line Hose, single jacket.......... **$131.00**
With double jacket hose.......... $155.00

No. E448. With Cable Cord Leader Line Hose, single jacket.......... **$135.00**
With double jacket hose.......... $159.00

WITH PROTECTOSEAL HOSE

Add $8.00 to the above prices if you want Underwriters, Riteflex or Cable Cord Hose with the new Protectoseal Treatment, described on Page 8.

☞ More Hose with the above outfits at prices shown at the right. Rocker Lug couplings, $2.92 per pair.

WE PAY FREIGHT ON LEADER LINE OUTFITS

3% Off for Cash with Order or C. O. D. Shipments

Champion

Guaranteed for 300-lb. test pressure.

No. C665. Champion Special Leader Line Hose, single jacket, per foot.......... **30c**

Double Jacket, guaranteed for 300-lb. test pressure, per foot.......... 45c

Underwriters

Guaranteed for 300-lb. pressure.

No. C668. Underwriters Leader Line Hose, single jacket, per foot.......... **40c**

Double Jacket, guaranteed for 400-lb. test pressure, per foot.......... 49c

No. C666P. Underwriters Leader Line Hose, WITH PROTECTOSEAL TREATMENT, single jacket, per foot.......... 44c

Double Jacket, per foot.......... 53c

Cable Cord

Guaranteed for 400 lb. pressure.

No. E445. Cable Cord Leader Line Hose, single jacket, per foot.......... **43c**

Double Jacket, guaranteed for 600-lb. test pressure, per foot.......... 55c

No. E445P. Cable Cord Leader Line Hose, WITH PROTECTOSEAL TREATMENT, single jacket, per foot.......... 47c

Double Jacket, per foot.......... 59c

Riteflex

Guaranteed for 400 lbs. pressure.

No. E446. Riteflex Leader Line Hose, single jacket, per foot.......... **41c**

Double jacket, guaranteed for 600 lbs. test pressure, per foot.......... 53c

No. E446P. Riteflex Leader Line Hose WITH PROTECTOSEAL TREATMENT, single jacket, per foot.......... 45c

Double jacket, per foot.......... 57c

America's Chiefs
Are Swinging to Champion

Hale, Waterous, Northern, Barton, Seagrave, Ahrens Fox, Howe, La France, Pirsch, Buffalo, Mack Pumps Accepted in Trade on Champion Midship and Front Pumps

No wonder Chiefs buy more Champions than other pumps. They know from their own tests and study that Champion is the finest.

Only Darley gives you so much for so little . . . so much quality for so little money . . . and that's why Champion's the Choice.

Furthermore! Darley gives a generous trade-in allowance for your old pump.

Because we know how hard it is for municipalities and Volunteer Fire Depts. to raise funds badly needed for modern equipment and to extend their service to rural areas—we invite any Fire Chief or Official to write us regarding trade-in allowance for any used fire pump to apply against purchase price of a new Champion, either Mid-Ship or Front.

Tell us the make, type, capacity, and year purchased and we'll make you the most liberal allowance.

Your fire truck will operate better than when new if you replace the old pump with a modern, high efficiency, centrifugal Champion.

You can secure a NEW centrifugal Champion for the cost of reconditioning your old pump; a few new parts won't help much—usually an old pump needs complete replacement of all working parts.

A Champion can easily be installed without chassis alterations.

CHAMPION CENTRIFUGALS REPLACE SCORES OF ROTARYS

Being a Fire Chief is an honorable and enjoyable career if you have dependable equipment. But if you have a badly worn pump of obsolete design, such as a rotary or piston, your life is a constant nightmare. Never knowing at what minute you'll get a call, you live in constant fear that the fire may be serious. Your pump, the heart of the fire truck, may still pump water, but you and your men know only too well that you can't depend on it. If it is a rotary type, it probably is suffering from slippage caused by sandy or gritty water. Sand or grit wears the smooth interior surfaces of a rotary like an abrasive and increases the clearances. The result is greater "slip" which causes loss of pressure and capacity.

Your old rotary will boost hydrant pressure but when you have to draft water your troubles begin. Precious time is lost when a few minutes makes the difference between a small blaze and a serious fire. Because of increasing clearances the old pump just won't pick up the water.

Unload your pumping troubles on us. We'll make it worth your while by giving a generous trade-in allowance.

Replace that old worn-out rotary or piston pump with a new Centrifugal Champion and your worries are over.

You'll thrill to the sense of power when you first operate a Champion. With effortless ease the Champion goes through the entire range of pressures and capacities. The Underwriters Approved, fully automatic primer is instantaneous in operation.

The Champion will always retain its initial high efficiency. Rotary fire pumps and centrifugals that depend on rotary primers cannot, because of their basic design and construction, retain priming efficiency, especially when subjected to pumping sandy or gritty water. The N. F. P. A. specifications now prohibit the use of rotary priming pumps for booster service.

For the kind of fire protection every Chief wants to give his community, replace your old pump with a Champion.

The Champion Centrifugal is an ideal replacement pump for every make of custom built apparatus. The illustration above shows the Champion Pump installed on a 5 Year Old Pirsch Engine in service at Park Ridge, Ill. No alterations required.

There Must Be a Reason Why Hundreds

SIMPLICITY OF DESIGN—Only one moving part, an impeller shaft supported on ball bearings outside the pump casing. Light weight. Less than weight of average man. No strain on chassis or on standard springs.

DEPENDABLE TROUBLEFREE PERFORMANCE—Complete absence of friction assures long, efficient life and permits us to make our Unconditional 10 Year Guarantee. One customer reports over 888 hours' continuous running, with the pump in perfect condition at the end of the run.

FLEXIBLE FRONT END DRIVE—Power drive between motor and pump is an axially free shaft of greater strength than motor crankshaft. Basically the same as the propeller shaft drive on all trucks. The only pump that may be safely installed on semi-floating or full-floating motors without disturbing original motor supports. Splined and riding on universal joints at each end, the shaft floats and compensates for any misalignment caused by flexing of chassis frame, thus preventing strain and damage to motor bearings.

DRIVE SHAFT—Drive shaft is splined and made of special chrome nickel heat-treated steel. Forged, then turned to nearly correct size, heat-treated and ground smooth. Shaft is short and adequately supported at each end by ball bearings so that it will never be deflected and sag in middle or wear.

CENTER OF GRAVITY—Center of gravity in the Champion is farther back and the pump is actually balanced over its mounting supports. Unlike other pumps, extending out and overhanging their mounting brackets, the Champion is not subjected to road shock, which causes serious clutch trouble and greatly affects efficiency. The extra supports used on such installations are merely temporary protection and require frequent adjustments if they are to be of any value whatsoever.

UNIVERSAL RIGID MOUNTING—Saddle mounting, 3 point support. The only pump adaptable to all makes and models of cars and trucks, old and new. Motor can be cranked the same as if no pump was installed. No special cranking device needed.

PUMP RUNS ONLY WHEN YOU ENGAGE IT—Totally disengaged except when pumping.

PHOSPHOR BRONZE IMPELLER—Enclosed type. Special design single seal ring type, for higher efficiency, longer life, easier maintenance.

BALANCED IMPELLER—Eliminates the possibility of a breakdown from end thrust on the bearings. Only engineers of long experience realize the terrific tons of pressure on the impeller and casing when a fire pump operates at high pressure.

In Champion design the impeller thrust is BALANCED—an exclusive feature not to be found in any other centrifugal pump. No danger of a cracked ball bearing; the very pressure of the water within the pump is diverted into slots so that thrust is practically eliminated.

IMPELLER SHAFT—Stainless steel, non-rusting. Shaft is short and adequately supported at both ends by ball bearings so that it will never be deflected and sag in middle or wear. Impeller is held in perfect alignment because no shaft deflection can occur.

PACKING PROTECTED FROM ABRASIVES—By means of centrifugal action all heavy, abrasive solids are separated from the water entering packing, which prevents scoring and excess wear of impeller shaft and increases life of packing.

BALL BEARINGS—Impeller shaft rides on ball bearings. Front bearings are double row shielded type. Rear bearings are deep groove type. Bearings have sealed-in lubrication. Double shielded against dirt and abrasives. All bearings are outside pump casing and are sealed from water.

TESTED TO 500 LBS. HYDRAULIC PRESSURE—Pump Casing is volute type. Semi-steel alloy castings. Modern metallurgy makes possible greater strength with lighter weight. (All bronze construction for handling salt water exclusively—small additional cost.)

AUTOMATIC PRIMER—Fully approved by the National Board of Fire, Underwriters and all State Insurance Bureaus. Completely automatic. The simple Champion system of priming consists of using the driving motor as a vacuum pump, combined with a positive closing automatic valve to prevent water being drawn into the motor.

BUILT-IN CHECK VALVE—A special Check Valve is built into the discharge head so that hose lines can be connected while pump is priming. Seals the discharge against entrance of air while priming but is opened automatically by water pressure. Non-clogging.

AUTOMATIC AUXILIARY MOTOR COOLING—Delivers a constant flow of clean water into the motor cooling system. A strainer in the cooling line protects the motor cooling system from foreign matter. The strainer cannot clog, constant centrifugal flow of water washes it clean. It is possible to pump for days without overheating motor.

PERFECT SEAL PACKING—The stuffing boxes at front and rear of main impeller shaft are adjustable type and only square graphite packing is required to insure perfect seal against air leakage. Pump Packing Glands very accessible.

LUBRICATION—Grease packed ball bearings. Lubrication sealed in. No other lubrication required.

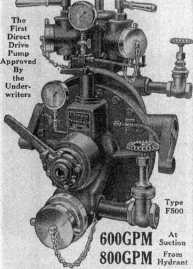

The First Direct Drive Pump Approved By the Underwriters

Type F500

600GPM At Suction

800GPM From Hydrant

SALIENT FEATURES OFFERED EXCLUSIVELY IN CHAMPIONS NOT FOUND IN ANY OTHER PUMP

FRICTION CLUTCH—Allows engagement and disengagement at varying motor speeds.

An exclusive Champion feature. Multiple disc, dry plate friction clutch. Built for a constant working load of 150 horsepower, 5 times greater than maximum load imposed upon it. Adjustment provision easily accessible at clutch lever. Many subjected to frequent, continuous, service for 5 years have never been adjusted.

FLOATING DRIVE—Insures trouble-free installation to all makes of trucks including semi-floating and full floating engines without disturbing original engine mounting.

POWER ASSEMBLY—Basically same as driving principle employed by truck manufacturers, i.e., an axially free shaft, riding on universals at front and rear. Front joint built in to front end of pump. Rear joint exposed between pump and motor. (A feature not possible when using common jaw or dog type clutches.)

CENTER BALANCE—Pump weight is balanced over sturdy mounting arms. Absolute freedom from road shocks and subsequent misalignment, an inherent weakness in projecting mounts employed in old type design with pump extending FULLY forward from mounting support.

UNIVERSAL MOUNTING—Saddle mounting. Three point support. Permits vertical and horizontal alignment when installation is being made so that thereafter never requires adjustment.

NO NEED for special adjustable front end support brackets. Adaptable to every make and model of car and truck chassis—bar none.

EASILY TRANSFERRED TO A NEW TRUCK AT NO ADDITIONAL EXPENSE

Perhaps yours is one of the many Depts. equipped only with Chemical Apparatus who are anxious but, because of finances, unable to purchase a complete NEW pumper. Give consideration to the installation of a FRONT MOUNT Champion, the Pump adaptable to any and every make of chassis. MOUNT it now to present equipment and later on when able convert it over to a new chassis at no extra cost. ONLY THE CHAMPION may be converted without costly alterations and investment of $100 to $150 for new mounting accessories.

Hundreds of Depts. have greatly increased the efficiency of their equipment by converting straight chemicals to triple combination Pumper, Booster, Hose and Ladder Trucks through the easy installation of a FRONT MOUNT CHAMPION.

MAKE A 30 DAY NO-RISK TEST – IF PUMP DOES NOT SATISFY

of Chiefs Say:
IT'S THE FINEST FRONT PUMP

$322

349

375

425

No. E715. 400 Gallon Underwriter Rated Capacity. Single 2½" Quick Action Discharge Valve. Combination Pressure and Vacuum Gauge. Plugged Booster Inlet and Outlet.

No. E716. 400 Gallon Underwriter Rated Capacity. Double 2½" Quick Action Discharge Valves. One Pressure Gauge and one Compound Pressure and Vacuum Gauge. Booster Inlet and Outlet Valves.

No. E717. 500 Gallon Underwriter Rated Capacity. Single 2½" Quick Action Discharge Valve. Combination Pressure and Vacuum Gauge. Plugged Booster Inlet and Outlet.

No. E718. 500 Gallon Underwriter Rated Capacity. Double 2½" Quick Action Discharge Valves. One Pressure Gauge and one Compound Pressure and Vacuum Gauge. Booster Inlet and Outlet Valves.

INCLUDED WITH EACH PUMP—Copper tubing and fittings for priming and cooling lines.

ALL Pumps Include

Built-In Check Valve
and
Quick Action Quarter Turn Discharge Valves
Features Not Available In Any Other Pump Selling Below $550

Meets Underwriter and State Bureau Tests

DIRECT SALES AT LOWEST PRICES

Guaranteed **10 YEARS** IN WRITING

But Built to Last a Lifetime!

U.S. GOVERNMENT SELECTS CHAMPION OVER ALL OTHERS

When the United States Government chooses Champion Apparatus with Front Mount Pump, you know it has the stamina and endurance that will give years of satisfactory service in your own department. Order your Champion Pump Now and make your own 30 Day Showdown Test.

We don't have any expensive factory representatives selling our pumps and you save their commission. We sell at low direct-from-factory-to-you prices with not one dollar of commission reserved for anyone.

Some pump makers find it necessary to sell through factory representatives, who visit your town and stay and talk, all at your expense. They don't install their pump on your chassis. After it's all done they simply come back and "check your whole installation—MAKE SURE EVERYTHING IS RIGHT."

Champions are not complicated. There is no need for installations by factory experts. Your local garageman can complete a highly satisfactory job of mounting without difficulty by following the simplified instructions that are included with the shipment—hundreds have done so.

The unnecessary expenses of high pressure salesmen and so-called "Factory Experts" for selling and mounting don't have to be added to Champion Prices. You pay no premium, yet secure the finest—from every angle—quality, simplicity, performance, when you select a Champion.

MAKE YOUR OWN SHOW DOWN TEST

We could talk for hours, and justifiably too, about Champion superiority. Always remembering, however, that performance speaks louder than words, we ask your permission to let the pump speak for itself, at absolutely no expense to you—no obligation whatever on your part. Request the shipment of a Champion Front Pump and use it in your Dept. under actual local conditions for 30 days.

Submit it to every test. "Give it the works" and compare it with any similar pump if you can prevail upon anyone else to ship on the same terms. Let the Pump do the talking and prove what it can do. Then if you don't agree with us that the Champion outperforms every other pump on the market (this includes pumps priced as high as $900.00), send it back at our expense.

Remember there is everything to gain and absolutely nothing to lose in giving first consideration to a Champion. Your chassis and motor are not mutilated in any way. Should the pump be disapproved for any reason whatever—any other unit may be installed in its place.

In your business dealings with Darley remember this: For over 30 years we have had only one aim . . . to illustrate and describe our equipment and service to you with absolute accuracy and truth, and by carrying this conviction to you, to win the privilege of serving you.

Our customers trust this Sales Book!

Owned by U. S. Bureau of Reclamation

RETURN IT—WE'LL PAY TRANSPORTATION CHARGES BOTH WAYS

Uncle Sam Places Official Approval on Champion Fire Trucks

U. S. Government, Panama Canal Zone, REORDERS!

During the latter part of 1936, a Champion Triple Combination was placed in service by the Panama Canal Purchasing Dept., Washington, D. C., at Gamboa, Panama. The service record of this engine, under operating conditions far more strenuous than any encountered here in the States, warranted the placement of additional orders for the two recent deliveries shown above.

The Government selected Champions only after competitive bids were received from other builders of Fire Apparatus. Government engineers drew up detailed specifications requiring Champion design and construction "or equal." But in the opinion of the impartial government engineers NO OTHER MAKER OF FIRE APPARATUS OR FIRE PUMPS offered the equal of the Champion.

W. S. DARLEY & CO. gave the U. S. Government the biggest value for its money, and government engineers recognized this fact or they would not have awarded us the two contracts.

It is a mistake to believe that the Government will buy only the lowest priced equipment, they can purchase higher priced equipment simply by drawing up their bid specifications to cover such equipment. BUT THE GOVERNMENT SPECIFIED CHAMPION OR EQUAL, which means that in their opinion THEY COULD NOT BUY BETTER FIRE APPARATUS FOR THIS SERVICE.

The Government Officials knew, from tests and performance in this country, how well Champions would serve in hardest kind of foreign service.

The Government Engineers specified Champion "or equal," and under "Service Requirements," they stated: "This fire truck will be put in service at the new townsite of Gamboa which is a very hilly district and which is over 10 miles from the station of the nearest assisting fire engine at Pedro Miguel. Because of this, the fire truck must be depended upon for continuous uninterrupted service and must be equipped with complete facilities for independent fire fighting operation."

Price has never been the consideration of Government Engineers in specifying fine fire fighting equipment. They recognize only merit, efficiency, performance and endurance as factors to be considered. The Government never assumes that the higher the price the greater the value. The judgment of the Government has been the findings of hundreds of Fire Chiefs everywhere who approve Champion Fire Pumps and Champion Fire Apparatus because they, too, know, FROM THEIR OWN TESTS of capacity and pressure, the Champion's actual worth as an investment in fire protection.

BUILT FOR U. S. GOV.
BUREAU OF
RECLAMATION

U. S. DEPARTMENT OF THE INTERIOR AWARDS CONTRACT TO DARLEY

Champion Fire Apparatus must be good to win the approval of a score of Gov. Engineers who inspected it at Wash., D. C. On a recent Gov. bid we quoted on Champion Apparatus of standard construction which we knew to be more than equal to specifications.

After the Gov. Engineers, who know Fire Apparatus from A to z, personally inspected Champion construction, they officially approved it. Their inspectors covered everything, with special attention to the Champion Centrifugal Pump. They checked conscientiously the engineering, design, construction, quality of materials, workmanship, arrangement of necessary equipment and finish. The official seal of approval on Champion Fire Apparatus was completed with the Award of the apparatus contract to DARLEY.

SOME OF MANY GOV. DEPTS. OWNING CHAMPIONS

Nat. Park Service, Grand Canyon, Ariz.
Bureau of Reclamation, Friant, Calif.
U. S. Mil. Reservation, Chillicothe, Ohio.
Navy Yard, Philadelphia, Pa.
Panama Canal Zone, Gamboa, Panama.
Navy Yard, Pearl Harbor, Hawaii.

Fort McHenry, Baltimore, Maryland.
Indian School, Albuquerque, N. M.
Naval Ammunition Depot, Fort Mifflin, Pa.
Naval Air Station, Cape May, N. J.
Dept. of Interior, Wash., D. C.

Champions Are Built Without the Burden of Unnecessary Weight

The Chief who owns a Champion is proud of it. So are we. They're well pleased with the way they perform, their fine appearance.

Note the pleasing proportions, and how the body nestles low on its frame, giving a low center of weight so essential to greater speed and ease of control; notice the initially mounted pump, the booster hose reel and equipment; how every part is accessible for quick action.

At a glance one gets the immediate impression of efficient power embodied in a correctly designed, well balanced unit built with precise knowledge of every requirement.

While this is but one example, it is typical of all Champion Apparatus and of how DARLEY Engineers plan the most scientifically designed apparatus in the world.

Champions are engineered for strength and represent more than just new eye-appeal. They are the first apparatus to eliminate the burden of unnecessary weight—weight that costs money to buy and more money to maintain and operate—weight that is as unnecessary in a modern fire apparatus as it is in a modern train or motor car.

Darley engineers, who in designing their new streamliners have successfully reduced the weight of Pullman cars from 70 tons to 43 tons. Darley engineers have designed a stronger yet lighter Fire Truck through smarter design and the use of the best materials available. They knew by eliminating useless weight they could cut costs and increase payload capacity, increase speed and pickup.

The average Champion Apparatus on a standard 1½-2 ton 157 inch wheel-base chassis weigh only about 4½ tons gross, including 200 gallons of water, 1,000 feet of 2½ inch fire hose and the usual equipment, such as booster and suction hose, extinguishers, tools, etc.

While the truck and its equipment include everything for effective fire fighting, it is not overloaded. Chiefs who command Champions say they handle like a passenger car. Because unsafe, dead weight has been eliminated, Champions powered by any of the commercial truck motors have acceleration and speed that gets men and equipment to the fire swiftly and safely.

While Champions are known and praised for road speed, it is where the going is toughest that they show their real superiority. Not overloaded and burdened by excess weight, Champions can go where big, heavy trucks would bog down. The Chief with a Champion can take water at suction wherever he can get his booster truck down close to a creek, into a farm yard, down to a river bank, even though the ground is soft. A Champion will pull in and OUT where other fire trucks could not approach, the going is toughest, have been selected time and time again by this government, foreign governments and municipal and industrial Fire Depts., for service here and abroad. Champions are serving faithfully all over the world.

When considering a new fire truck, you can accept these beautiful, power-ful, fast, dependable Champions—either on faith, based on the character and reputation of S. S. Darley & Co. in business for over 30 years. Or on a careful study of their modern engineering. Or, on both, keeping in mind that Champions are America's fastest selling Fire Apparatus. Write Chiefs who command Champions, let them tell you in their own words, man-to-man, about Champion's superior performance and stamina.

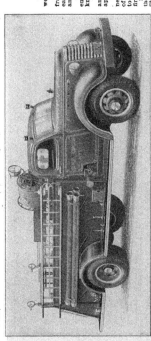

GMC—BUILT FOR HICKORY TOWNSHIP, SHARON, PA.

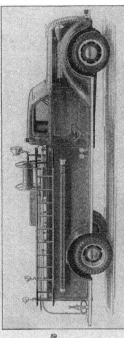

DIAMOND T—BUILT FOR WESTFIELD, N.Y.

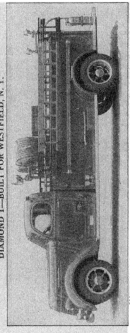

INTERNATIONAL—BUILT FOR INTERNATIONAL HARVESTER CO.

CHIEF DUNCAN SAYS:
"Champion Is the Equipment of MODERN Depts."

W. S. DARLEY & CO.
Chicago, Illinois.
Gentlemen:

Our Champion Pumper has been thoroughly efficient and we are more than satisfied with the results of the work done with it.

The various trucks have set many records of fire fighting and quick response that have never been equalled by any other company in our department.

The water of tremendous capacity, and smashing pressure obtained from the CHAMPION pump is indeed astounding. Truly, CHAMPION is "the equipment of modern fire departments" as every official must admit. The pump has never commanded such respect.

Please accept our most sincere appreciation to your worthy firm for the creation of the masterpiece of fire fighting apparatus ... "CHAMPION."

Very truly yours,
J. F. DUNCAN, Jr.
Assistant-Chief,
Beaufort, N. C., Fire Dept.

Standard $1342

CHAMPION

CONSTRUCTION YOU CAN'T BUY ANYWHERE NEAR OUR LOW PRICES

All Apparatus prices are for Champion Basic Equipment built on truck chassis you supply through your local dealer. Or we can supply any chassis at the delivered factory price and save you all the transportation charges to Chicago. We will take delivery of the chassis we purchase for you at the factory. Detroit for Ford, Dodge and Chevrolet. Pontiac, Mich., for GMC. Fort Wayne, Ind., for International: Chicago for Diamond T. 1½ to 3 ton, conventional or cab over engine model. With windshield only or with cab. Any standard wheelbase such as 157 inches.

TYPE — TRIPLE COMBINATION, PUMPER, BOOSTER, HOSE AND LADDER APPARATUS. Design, construction and performance fully approved by National Board of Fire Underwriters, the highest authorities on Fire Apparatus. Accepted by Insurance Bureaus in all States as equal to the most expensive apparatus.

BODY—Ruggedly constructed of 12 gauge copper bearing steel. One-piece body walls extend down to running boards and rear platform step and are rigidly supported and braced by welded square tubular uprights at ends. Streamlined. Inside length of body on 157 inch wheelbase chassis, approximately 11 feet. Inside height 27 to 30 inches, depending on load requirements. To insure solid foundation body is supported full length by steel extensions electrically welded to chassis frame and formed of same steel used in chassis. Built to withstand extreme distortion of chassis, vibration and road shocks.

Hose Capacity 1400 feet, 2½ inch regulation double jacket, cotton, rubber lined, fire hose. Body has smooth interior free from all sharp projections, which might injure hose.

SPECIAL COMPARTMENT—Very roomy. Built-in, back of seat.

Price list

$1452

STANDARD TYPE M500
Champion Fire Apparatus built to standard specifications with Type M500 Gear Drive Centrifugal Midship Pump. Underwriter Approved for 500 gallon rating on Mack 5 M.P., Reo, Mack, White, Federal, Stewart, etc.

$1372

STANDARD TYPE M400
Champion Fire Apparatus built to standard specifications with Type M400 Gear Drive Centrifugal Midship Pump. Underwriter Approved for 400 gallon rating on Ford 5¼ and 95 H.P., Chevrolet, International, GMC, Diamond T, Studebaker, Reo, Mack, White, Federal, Stewart, etc.

$1342

STANDARD TYPE MF500
Champion Fire Apparatus built to standard specifications with Type MF500 Direct Drive Centrifugal Midship Pump. Underwriter Approved for 500 gallon rating on Ford 95 H.P. chassis.

$1277

STANDARD TYPE MF400
Champion Fire Apparatus built to standard specifications with Type MF400 Direct Drive Centrifugal Midship Pump. Underwriter Approved for 400 gallon rating on Ford 5¼ and 95 H.P., Chevrolet, Dodge, International, GMC, Diamond T, etc.

$1204

STANDARD TYPE F500
Champion Fire Apparatus built to standard modifications with Type F500 Front End Centrifugal Pump. Underwriter Approved for 500 gallon rating on Ford 95 H.P. chassis.

$1128

STANDARD TYPE F400
Champion Fire Apparatus built to standard specifications with Type F400 Front End Centrifugal Pump. Underwriter Approved for 400 gallon rating on Ford 95 and 95 H.P., Chevrolet, Dodge, International, GMC, Diamond T, etc.

60

LADDER BRACKETS—Elongated U Shape, quick action type. All-steel, welded to both sides of body.

LADDER LOCKS—Spring loaded eccentric type, clamping over the rungs at each end of the ladder. Hold the ladder securely, preventing chafing.
Ladder easily removed by releasing the quick action locking clamps and pulling ladder horizontally away from the side of the truck—it being unnecessary to lift the ladder or more it in any other direction.
Two ladder brackets and two locks on each side of body. For carrying an extension ladder, two section or three section type, and a wall and roof ladder, any standard length. Or the brackets and locks can be made to carry an extension ladder and a roof ladder on each side of the truck. Brackets and locks can be constructed to carry any ladders, new or old.

BOOSTER TANK—Rust-resistant copper bearing steel. Mounted forward in hose body, entirely separate from body, not a part of the body. Capacity 100 gallons or built to your specifications. (Copper tank furnished on special order.)
Welded construction. Baffle plates prevent water from sloshing. Top depressed 4 inches to make easy filling trough extending full width and 24 inches of tank length. Water dumped into trough from either side drains into tank through 7 inch center opening which has removable screen and quick action slide cover.

BOOSTER INLET CONNECTIONS—Connections from tank to pump through pipe fittings and extra heavy non-collapsible rubber hose. Shutoff valve at tank outlet is easily accessible on driver's side.

BLEEDER VALVE IN BOOSTER INLET LINE—Drains line dry to protect against freezing.

BOOSTER DISCHARGE CONNECTIONS—Special line for high pressure discharge from pump to booster hose reel. Connections from pump to reel through pipe fittings and extra heavy rubber hose that stands test pressure of 500 lbs. Shutoff valve at pump outlet for shutting off the line to reel.

BOOSTER HOSE BASKET—All-steel. Perforated ends for ventilation. Smooth interior construction. Capacity 250 feet of ¾ inch hose, or 150 feet of 1 inch hose.
SEAT—Extra wide, spring built, custom built. Removable, heavily upholstered cushions. Covered with specially fabricated waterproof and weatherproof material, far more durable than leather.

Special Type

61

REAR PLATFORM STEP—All-steel construction, with top surface of corrugated rubber for secure footing. Bound in aluminum. Extends full width of fenders and 18 inches out from rear of body. Average height above ground is 20 inches.
Not supported by body, made an integral part of chassis by means of channel construction, welded into cross sill.

RUNNING BOARDS—All-steel construction, with top surface of corrugated rubber for secure footing. Bound in aluminum. Supported and braced at intervals by angle frame-work welded to chassis. Provide generous space for carrying full equipment.
FENDERS—Specially fabricated modern type. Cover entire apparatus. Crown as to running boards and rear platform step. Rolled from heavy gauge steel, welded reinforced construction, strongly braced to withstand vibration. Joined to running boards and rear platform step.

RAILINGS—Rear handrail and assist rails on each side of step. Heavily chromium plated, 1¼ inch diameter brass. Cirumed brass mounting handles on each windshield stanchion.

SUCTION HOSE CARRIER—Fabricated from 12 gauge copper bearing steel. Carries two or three 10 foot lengths of hose. Welded to right side of body.

SPECIAL THROTTLE—Quadrant type friction lock throttle. For Front Pump throttle is installed on radiator shell. For Midship Pump throttle is installed within easy reach from both sides of truck.

AUXILIARY TOOL BOX—All-steel rugged construction. Size, 8x8x21 inches.
POWERFUL SEARCHLIGHT—Special Fire Apparatus type, big 10 inch diameter, largest size. Gives long range searchlight beam. Heavily chromium plated. Swivel mounted so as to swing up or down and in any direction.
HOSE LIGHTS—Two, 235,000 candle power. Largest size, 6½" diameter. All brass. Chromium plated. Switch built into each light. Swivel mounted so as to swing up or down and in any direction.

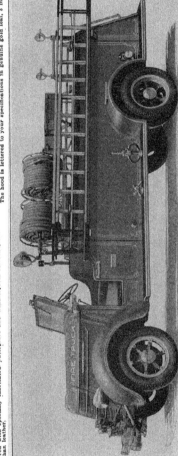

COMBINATION TAIL LIGHT AND REAR STEP LIGHT—Mounted at left side of body below floor level, just above rear step. Throws white light to fully illuminate the entire step.
BUILT-IN RED LIGHT OVER GAUGES—Chromium plated, weatherproof, with red light which throws light directly on gauges. Standard on all Midship Pumper Apparatus.
ELECTRIC SIREN—Super Champion High Speed Fire Apparatus type. Heavily chromium plated. Convenient foot push switch.
GAUGES—Two, special heavy duty Fire Apparatus type. One Pressure Gauge and one Combination Pressure and Vacuum Gauge.

TACHOMETER—For indicating motor-pump speed.

FINISH, STRIPING, LETTERING—The entire Fire Apparatus, including superstructure, chassis, wheels and running gear is handsomely finished in Darley Fire Engine Red Enamel. Fire chassis are custom built, durable and long wearing under varied climatic exposure and hard Fire Dept. service.
Beautiful gold striping, panels and corner designs.
The hood is lettered to your specifications in genuine gold leaf, 3 inch, block letters.

STANDARD and DELUXE APPARATUS BUILT ON ANY TRUCK CHASSIS YOU SELECT

DeLuxe CHAMPION $1437

WHEN COMPARING PRICES, CHECK SPECIFICATIONS TO PROVE CHAMPION IS THE GREATEST VALUE

All Apparatus prices are for Champion Basic Equipment built on truck chassis you supply through your local dealer. Or we can supply any chassis at the delivered factory price and save you all the transportation charges to Chicago. We will take delivery of the chassis we purchase for you at the factory. Detroit for Ford, Dodge and Chevrolet; Pontiac, Mich., for GMC; Fort Wayne, Ind., for International; Chicago for Diamond T.

Chassis may be any 1½ to 3 ton, conventional or cab over engine model. With windshield only or with cab. Any standard wheelbase such as 157 inches.

TYPE—TRIPLE COMBINATION, PUMPER, BOOSTER, HOSE AND LADDER APPARATUS. Design, construction and performance fully approved by National Board of Fire Underwriters, the highest authorities on Fire Apparatus. Accepted by Insurance Bureaus in all States as equal to the most expensive apparatus.

BODY—Ruggedly constructed of 12 gauge copper bearing steel. One-piece body walls extend down to running boards and rear platform step and are rigidly supported and braced by welded square tubular uprights at ends. Streamlined. Inside length of body on 157 inch wheelbase chassis, approximately 11 feet. Inside height 27 to 30 inches, depending on load requirements. To insure solid foundation, body is supported throughout full length by steel extension channels welded to chassis frame and formed of same steel used in chassis. Built to withstand extreme distortion of chassis, vibration and road shocks.

Hose Capacity 1400 feet 2½ inch regulation double jacket, cotton, rubber lined, fire hose. Body has smooth interior free from all sharp projections, which might injure hose.

SPECIAL HOSE BODY FLOORING—Double flooring with top flooring slatted for air circulation around hose to prevent mildew.

DELUXE TYPE M500 $1547

Champion Fire Apparatus built to Deluxe specifications, with Type M500 Gear Drive Centrifugal Midship Pump, Underwriter Approved for 500 gallon rating, on Ford 85 and 95 H.P. Chevrolet, Dodge, International, GMC, Diamond T, Studebaker, Reo, Mack, White, Federal, Stewart, etc.....

DELUXE TYPE M400 $1467

Champion Fire Apparatus built to Deluxe specifications, with Type M400 Gear Drive Centrifugal Midship Pump, Underwriter Approved for 400 gallon rating, on Ford 85 and 95 H.P., Chevrolet, Dodge, International, GMC, Diamond T, Studebaker, Reo, Mack, White, Federal, Stewart, etc.

DELUXE TYPE MF500 $1437

Champion Fire Apparatus built to Deluxe specifications, with Type MF500 Direct Drive Centrifugal Midship Pump, Underwriter Approved for 500 gallon rating on Ford 85 H.P. chassis....

DELUXE TYPE MF400 $1377

Champion Fire Apparatus built to Deluxe specifications, with Type MF400 Direct Drive Centrifugal Midship Pump, Underwriter Approved for 400 gallon rating on Ford 85 and 95 H.P., Chevrolet, Dodge, International, GMC, Diamond T, etc.....

DELUXE TYPE F500 $1285

Champion Fire Apparatus built to Deluxe specifications, with Type F500 Direct Drive Centrifugal Front Pump, Underwriter Approved for 500 gallon rating on Ford 85 H.P. chassis.....

DELUXE TYPE F400 $1209

Champion Fire Apparatus built to Deluxe specifications, with Type F400 Direct Drive Centrifugal Front Pump, Underwriter Approved for 400 gallon rating on Ford 85 and 95 H.P., Chevrolet, Dodge, International, GMC, Diamond T, etc.....

SPECIAL COMPARTMENTS — One compartment is in rear of body. All-steel. Heavy steel door, mounted on durable piano style hinges. Spring lock. For tools.

Another roomy compartment in back of seat.

LADDER BRACKETS—Elongated U Shape, quick action type. All-steel, welded to both sides of body.

LADDER LOCKS—Spring loaded eccentric type, clamping over the rungs at each end of the ladder. Hold the ladder securely, preventing chafing.

Ladder easily removed by releasing the quick action locking clamps and pulling ladder horizontally away from the side of the truck—it being unnecessary to lift the ladder or move it in any other direction.

Two ladder brackets and two locks on each side of body. For carrying extension ladder, two section or three section type and a wall and roof ladder, any standard length. Or the brackets and locks can be made to carry an extension ladder and a roof ladder on each side of the truck. Brackets and locks can be constructed to carry any ladders, new or old.

BOOSTER TANK—Rust-resistant copper bearing steel. Mounted forward in the body but not a part of the body. Capacity 100 gallons or built to your specifications. (Copper Tank furnished on special order.)

Welded construction. Baffle plates prevent water from sloshing. Top depressed 4 inches to make easy filling, trough extending full width and 24 inches of tank length. Water dumped into trough from either side drains into tank through 7 inch center opening which has removable screen and quick action slide cover.

BOOSTER INLET CONNECTIONS—Connections from tank to pump through pipe fittings and extra heavy non-collapsible rubber hose. Shutoff valve at tank outlet is easily accessible on driver's side.

Shutoff valve at pump inlet for shutting off the line from tank when pumping from suction or hydrant.

BLEEDER VALVE IN BOOSTER INLET LINE—Drains line dry to protect against freezing.

BOOSTER DISCHARGE CONNECTIONS—Special line for high pressure discharge from pump to booster hose reel. Connections from pump to reel through pipe fittings and extra heavy rubber hose that stands test pressure of 600 lbs. Shutoff valve at pump outlet for shutting water off line.

AUTOMATIC BOOSTER HOSE REEL—Mounted on welded steel supports over tank. Permanently connected to pump, always ready for instant use. Capacity, 200 feet 1 inch hose. Crank handle for winding hose. Mechanical rear installation under body is optional.

SEAT—Extra wide cushion, full spring cushioned. Removable, heavily upholstered, far more durable than leather. Covered with specially fabricated waterproof and weatherproof material.

REAR PLATFORM STEP—All-steel construction, with top surface of corrugated rubber for secure footing. Bound in aluminum. Extends full width of fenders and 18 inches

out from rear of body. Average height above ground is 20 inches. Not supported by body, made an integral part of chassis by means of channel construction, welded into one unit, and braced at intervals by angle frame-secure footing. Bound in aluminum. Supported and braced at intervals by angle framework, welded to chassis for carrying full equipment.

RUNNING BOARDS—All-steel construction, with top surface of corrugated rubber for secure footing. Bound in aluminum. Supported and braced at intervals by angle framework, welded to chassis for carrying full equipment.

FENDERS—Specially fabricated for each Apparatus. Crown type. Rolled from heavy gauge steel, welded reinforced construction, strongly braced to withstand vibration.

FENDER STEP PLATES.—Cast aluminum, highly polished trim, corrugated rough tread pattern given secure and safe footing. Size 8x11 inches.

RAILINGS—Streamlined pattern, heavily chromium plated. Front, side, rear cross bar and step railings are 1¼ inch diameter.

Chromeplated brass mounting handles on each side of seat and on each windshield stanchion.

HANDRAILS—Streamlined, fabricated from 1¼ gauge copper bearing steel. Carries two or three 10 foot lengths of hose. Welded to right side of body.

SPECIAL TAIL LIGHT AND STOP.—Extra large combination, Two-plate steel, will never seals or rust. Securely mounted in body between front body wall and booster tank. Fuller opening outside body. Made to Underwriter specifications, capacity 22 gallons.

SPECIAL THROTTLE—Quadrant type Friction lock throttle. For Front Pump throttle is installed on radiator shell. For Midship Pump throttle is installed within easy reach from both sides of truck.

AUXILIARY TOOL BOX—All-steel rugged construction. Size, 8x6x21 inches.

POWERFUL SEARCHLIGHT—Special Fire Apparatus type, big, 10 inch diameter, largest size. Gives long range searchlight beam. Heavily Chromium Plated. Swivel mounted so as to swing up or down and in any direction.

HOSE LIGHT.—Two, 32,000 candle power. Largest size, 6½ inch diameter. All brass Chromium plated. Switch built into each light. Swivel mounted so as to swing up or down and in any direction.

COMBINATION TAIL LIGHT AND REAR STEP LIGHT.—Mounted at left side of body, below floor level just above rear step. Throws white light to fully illuminate the entire step.

GAUGES—Four, special heavy duty Fire Apparatus type. Two Pressure Gauges, one each side. Two Combination Pressure and Vacuum Gauges.

BUILT-IN LIGHTS OVER GAUGES.—Chromium plated, weatherproof, with special reflector which throws light directly on gauges. Standard on all Midship Pumper Apparatus.

ELECTRIC FLASHER SIREN.—Streamlined Chromium Plated Siren with brilliant red light that flashes 50 times a minute. Largest size. Convenient foot push switch.

TACHOMETER—For indicating motor-pump speed.

FINISH, STRIPING, LETTERING.—The entire Fire Apparatus, including super-structure, chassis, wheels and running gear is hand-spray painted. Rear installation under body is optional. First class custom finish, durable and long wearing under varied climatic exposure and hard Fire Dept. service.

Beautiful gold striping, panels and corner designs.

The hood is lettered to your specifications in genuine gold leaf, 3 inch block letters.

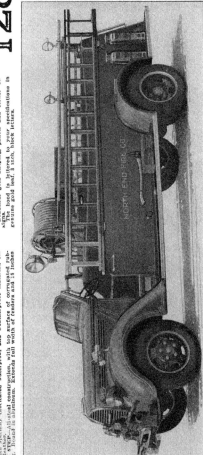

NORTH END FIRE CO.

$1285

63

The Most Efficient Fire Pump

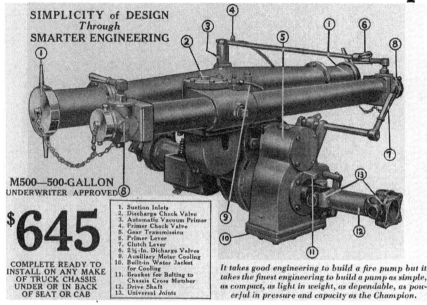

SIMPLICITY of DESIGN
Through
SMARTER ENGINEERING

M500—500-GALLON
UNDERWRITER APPROVED

$645

COMPLETE READY TO
INSTALL ON ANY MAKE
OF TRUCK CHASSIS
UNDER OR IN BACK
OF SEAT OR CAB

1. Suction Inlets
2. Discharge Check Valve
3. Automatic Vacuum Primer
4. Primer Check Valve
5. Gear Transmission
6. Primer Lever
7. Clutch Lever
8. 2½-In. Dicharge Valves
9. Auxiliary Motor Cooling
10. Built-in Water Jacket for Cooling
11. Bracket for Bolting to Chassis Cross Member
12. Drive Shaft
13. Universal Joints

It takes good engineering to build a fire pump but it takes the finest engineering to build a pump as simple, as compact, as light in weight, as dependable, as powerful in pressure and capacity as the Champion.

EASILY THE BEST PUMP *No matter how you compare them*

No wonder we invite comparison when our Champion Fire Pumps are engineered and built finest, yet priced lowest.

Champions are centrifugals, of course, the most popular type with Fire Chiefs the country over. A reliable survey made recently showed that over 75% of all fire pumps being sold are centrifugal.

Champions are superior not only because they are centrifugal pumps but because they are the simplest and most efficient pumps yet designed for modern, commercial truck chassis. Not adapted from pumps originally designed for slow speed apparatus chassis. Champion pumps had their exclusive origin on the drawing boards of Darley engineers, who had nothing to go by in other pumps because they were out to design a new pump special for commercial chassis.

How well they succeeded is best told by the splendid performance records of the hundreds of Champions in service all over the world. Fire Chiefs who command Champions can best testify to Champions performance and they do with all the enthusiasm of men who know a good thing when they see it.

Champion performance is all the more astonishing in view of the simplicity of construction and operation. In Champions you get modern engineering skill, which makes fewer parts serve the purpose of the complicated assemblies found in other pumps. A good engineer can make a piece of mechanical equipment work with a certain number of parts. But a better engineer can make the same piece of equipment work with fewer parts.

Because the Champion is better engineered hydraulically and mechanically it uses only two gears. No train of gears as found in other pumps. Not having a series of grinding gears, it seems, to

watch a Champion centrifugal in action, that vibration has vanished. Beware when gearing shouts in a pump. In a train of gears noise is the voice that shouts "Here is vibration—here is wear—here is a weakness that shortens the life of a pump, interferes with sustained high performance, adds to maintenance and replacement costs."

The Champion centrifugal has no complicated power takeoff rotary pump for priming. When you place a Champion in service you get every advantage of a centrifugal pump without the handicap of depending on a miniature rotary pump to prime it. If you are going to depend on a rotary pump to prime the centrifugal you might as well buy a rotary fire pump in the first place. Gritty water causes wear and greater clearances in a small rotary just as quickly as in a large rotary. Increased clearances in a rotary pump means greater slip (leakage past the rotors). Since slip represents loss of pressure and capacity, more power and speed must be added to maintain normal performance as the pump clearance is opened up by gritty water. But nothing can be done to make a wornout rotary pump prime. And any centrifugal that depends on a rotary to prime it is just as helpless.

The simple Champion system of priming consists of using the driving motor as a vacuum pump, combined with a positive closing automatic valve to prevent water being drawn into the motor. A special, non-clogging check valve seals the discharge against entrance of air while priming. Our Automatic Vacuum Primer is fully approved by the National Board of Fire Underwriters and all State Insurance Bureaus.

These are only a few of the features that make Champion the best buy in a fire pump—and the most for your money.

UNDERWRITER APPROVED 500-GALLON MIDSHIP

Ever Built for Midship Mounting

SIMPLICITY OF DESIGN—Weight, only 650 lbs., about half the weight of other pumps. Modern engineering coupled with sound knowledge of metallurgy result in less weight (all dead weight eliminated) and even greater rugged strength than other pumps of obsolete design. Every casting and part is subjected to hydraulic pressures greatly in excess of Underwriter requirements. Safety factors are very great. The simplest Fire Pump made for mounting midship. Only one moving part—an impeller shaft supported on ball bearings outside the pump casing.

DEPENDABLE PERFORMANCE—Complete absence of friction because of this simple construction, assures long, efficient life with an Unconditional 10-Year Guarantee.

GEAR DRIVE OFF PROPELLER SHAFT—The only Midship Fire Pump without complicated drive mechanism. Drives through gears off propeller shaft from motor. Only two gears, the finest possible to produce. Simplicity insures freedom from mechanical failure.

Installation of the pump has no effect whatever on the normal operation of any truck. Power and speed for propelling the truck are not changed. No part of chassis, motor, transmission or propeller shaft is mutilated or weakened.

HERRINGBONE GEARS—Because of built-in efficiency higher than ever before achieved in a fire pump, ONLY 2 gears are required to deliver the Underwriter tests with ease. Powered by any truck motor, the Type M500 actually delivers 600 GPM at suction and 800 GPM from hydrants. The Type M400 is Underwriter rated at 400 GPM but actually delivers 500 GPM at suction and 650 GPM from hydrants.

Motor operates at about half pump speed.

Gears are, continuous tooth herringbone type, made of highest grade alloy steel, heat-treated, running in oil bath. The finest gears money can buy. Far superior to all other types such as spur and helical gears. Quieter, smoother running, stronger and no end thrust.

In keeping with sound engineering practice, the gears are built with axial freedom. Some pumps are inherently defective in this respect; a serious defect because in every set of herringbone gears one gear must be free to shift into correct alignment with its mate at all times.

The single set of gears is enclosed in a steel gear box which has a built-in water jacket as an extra cooling feature. Gears are sealed from water. Continuous water circulation keeps the oil at an even, cool temperature, and prevents overheating when pumping for many hours or even days at a time.

The gear Ratio is from 2-1 to 2½-1 and varies for different motors.

DRIVE SHAFTS—Chrome nickel steel, heat-treated. Finest possible to produce. Ground to truck chassis manufacturer's specifications. Forged, then turned to nearly correct size, heat-treated and ground smooth. Polished to mirror-like finish where ball bearings fit and at other essential points. Calibrated when finished to fit bearings to within one five-thousandths of an inch after polishing. Very rigid and very strong. Shafts are short and adequately supported at each end by ball bearings so that they will never be deflected and sag in the middle or wear. Even in their smallest section the drive shafts are stronger than the standard truck shaft.

UNIVERSAL JOINTS—Built into the pump at both front and rear are universal joints exactly the same as the two universal joints used in every truck chassis for connecting the short drive shaft to the transmission and to the propeller shaft of the chassis. The pump is mounted on the chassis frame so that its drive shafts and universal joints set in place of the short coupling shaft and universal joints which are standard on all truck chassis.

Pump is securely anchored in position by bolting through frame on each side of chassis. Special brackets front and rear of pump for bolting to chassis cross member and also to special cross member.

Pump is rigidly mounted on chassis frame to ride safely without possible injury to pump or chassis.

SELECTIVE CLUTCH—The clutch is without compare for simplicity. Sliding splined sleeve type—selective engagement of pump or propeller shaft. Chrome nickel steel, heat-treated. Perfectly lubricated. Operated by lever from driver's seat. Positive safety locking device assures full engagement and full disengagement and prevents accidental shifting of clutch lever.

PUMP RUNS ONLY WHEN YOU ENGAGE IT—When you are driving to a fire no pump parts are moving, the pump is totally disengaged.

IMPELLER SHAFT—Stainless steel, non-rusting. Shaft is short and adequately supported at two points by ball bearings so that it

(Continued on next page)

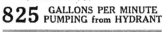

825 GALLONS PER MINUTE PUMPING from HYDRANT

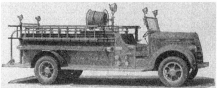

Guaranteed
10 YEARS IN WRITING

If for any reason whatsoever, at any time within the 10 Year Guarantee Period, your Champion Fire Pump fails to give satisfactory service, we will replace, free of charge, any defective part or the entire pump.

W. S. DARLEY & CO.

Superior Hydraulic and Mechanical Design Gives Champion Higher Efficiency Than Any Other Pump. Only a Champion Can Deliver the Underwriter's 500-Gallon Test on the D30 International Chassis with 72-Horse Power Motor.

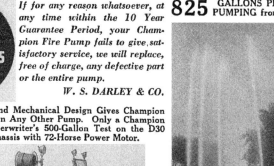

CENTRIFUGAL PUMP PRICED TO SAVE YOU $500

will never be deflected and sag in middle or wear. Impeller is held in perfect alignment because no shaft deflection can occur.

PHOSPHOR BRONZE IMPELLER—Enclosed type impeller of phosphor bronze with single seal ring for higher efficiency, longer life, easier maintenance. All other pumps have two seal rings and much greater wearing surface.

BALANCED IMPELLER—Eliminates the possibility of a breakdown from end thrust on bearings. Only engineers of long experience realize the terrific tons of pressure on the impeller and casing when a fire pump operates at high pressure.

In Champion design the impeller thrust is BALANCED—an exclusive feature not in any other centrifugal pump. No danger of a cracked ball bearing; the very pressure of the water within the pump is diverted into slots so that end thrust is practically eliminated.

PACKING PROTECTED FROM ABRASIVES—By means of centrifugal action all heavy, abrasive solids are separated from the water entering packing, which prevents scoring and excess wear of impeller shaft and increases life of packing.

BALL BEARINGS—Adequate bearings throughout. Deep groove and self aligning type ball bearings with extra load capacity for very great safety factor. Double shielded against dirt and abrasive. Perfectly lubricated.

No submerged or water lubricated bearings. All bearings are outside pump casing and sealed from water.

TESTED TO 500 LBS. HYDRAULIC PRESSURE—Pump Casing is volute type. Semi-steel alloy castings. Modern metallurgy makes possible greater strength with lighter weight. (At small extra cost pumps can be made all-bronze for salt water pumping service.)

AUTOMATIC PRIMER—Fully approved by the National Board of Fire Underwriters and all State Insurance Bureaus. Completely automatic. The simple Champion system of priming consists of using the driving motor as a vacuum pump, combined with a positive closing automatic valve to prevent water being drawn into the motor.

BUILT-IN CHECK VALVE—A special Check Valve is built into the discharge head so that hose lines can be connected while pump is priming. Seals the discharge against entrance of air while priming but is opened automatically by water pressure. Non-clogging.

AUTOMATIC AUXILIARY MOTOR COOLING—Delivers a constant flow of clean water into the motor cooling system. A strainer in the cooling line protects the motor cooling system from foreign matter. The strainer cannot clog, constant centrifugal flow of water washes it clean. It is possible to pump for days without overheating motor. (See Page 16 for self-contained cooling system.)

PERFECT SEAL PACKING—The stuffing box of main impeller shaft is adjustable type and only square graphite packing is required to insure perfect seal against air leakage at all times. Pump Packing Gland easily reached and readily accessible.

LUBRICATION—All bearings and transmission parts run in oil.

SPECIAL EMERGENCY BRAKE—For all chassis with emergency brake on the propeller shaft we build a special emergency brake into the pump assembly. Controlled by lever from driver's seat.

TWO 2½-INCH DISCHARGE OUTLET VALVES—Quick Action Quarter Turn Shutoff Type, all bronze.

BOOSTER INLET SHUTOFF VALVE—For shutting off the line from tank when pumping from suction or hydrant.

BOOSTER DISCHARGE SHUTOFF VALVE—For shutting off the line to reel when booster hose is not in use.

BRASS DRAIN COCK—Drains pump absolutely dry and provides full protection against freezing in cold weather.

SUCTION INLETS—Size 4 inch on the M500 Pump. Size 3½ inch on the M400 Pump. Built-in strainers for positive protection when pumping from hydrants.

GAUGES—Four, special heavy duty type, for Fire Apparatus. Two pressure gauges calibrated 0-300 lbs. Two compound gauges calibrated 0-30 inches and 0-300 lbs.

THE ONLY PUMP DESIGNED FOR COMMERCIAL CHASSIS

Easily Delivers Underwriter Tests

Efficiency of Design and Performance No Other Pump Can Equal

MAKE THIS 30-DAY "SHOW-DOWN TEST"

We invite you to PROVE to YOURSELF in your own town every word we have said about the wonderful Champion Fire Pump. Thorough TESTS will convince you it can't be beaten as a fire fighter. Compare its mechanical construction with the highest priced fire pumps.

Order the pump you desire in the regular way. Make every Fair Test you wish for a full 30 Days and satisfy yourself that you've saved over 50%. Then, if you are not satisfied, return it to us and we'll refund the money you have paid, including all transportation charges. Could any offer be fairer? You don't risk a cent, but you will make a BIG SAVING.

Type M500 Champion Gear Drive Centrifugal Midship Fire Pump, Underwriter Rated 500 GPM, actually delivers 600 GPM from suction and 800 GPM from fair hydrants, complete..................**$645**

Type M400 Champion Gear Drive Centrifugal Midship Fire Pump, Underwriter Rated 400 GPM, actually delivers 500 GPM from suction and 650 GPM from fair hydrants, complete..............**$565**

When Ordering Tell Us Make of Chassis, Model, Year Built.

EASILY INSTALLED ON ANY FIRE APPARATUS or CHASSIS, Old or New

The installation is very easily made without alterations to motor or chassis. Built into the pump at both front and rear are universal joints exactly the same as the two now used connecting the short drive shaft to the transmission and propeller shaft of the chassis. After removing this drive shaft (only a wrench is needed for the job) the pump slips into position; having been engineered for your truck there is no trick or trouble to get the proper location —it just has to rest in the right position. Make your universal connections at both front and rear and the job is done except for drilling two holes (one on each side of frame) through which bolts are passed for anchoring the pump in position.

Our low price includes everything as described and illustrated. Everything needed to make a complete installation. With each pump we send Easy Blue Prints and Easy Instructions for mounting.

You Can Install a Pump On Any Chassis In a Few Hours

Why Pay $1000 or More? This Pump Is Underwriter Approved and Costs Only $645 or $625⁶⁵ Cash with Order or C.O.D.

CLASSES OF SERVICE

DOMESTIC SERVICES	CABLE SERVICES

TELEGRAMS
Full-rate expedited service.

ORDINARIES
The standard service, at full rates. Code messages, consisting of 5-letter groups only, at a lower rate.

DAY LETTERS
Deferred service at lower than the standard telegram rates.

SERIALS
Messages sent in sections during the same day.

DEFERREDS
Plain-language messages, subject to being deferred in favor of full-rate messages.

NIGHT LETTERS
Accepted up to 2 A.M. for delivery not earlier than the following morning at rates still lower than the standard telegram or day letter rates.

NIGHT LETTERS
Overnight plain-language messages.

SHIP RADIOGRAMS
Service to ships at sea, in all parts of the world. Plain language or code language may be used.

URGENTS
Messages taking precedence over all other messages except government messages.

THERE IS A SPECIAL LOW-RATE WESTERN UNION SERVICE FOR EVERY SOCIAL NEED

Telegrams of the categories listed at the right, to any Western Union destination in the United States

GREETINGS AT
Christmas, New Year, Easter
Valentine's Day, Mother's Day, Father's Day
Jewish New Year, Thanksgiving

TELEGRAMS OF PRESCRIBED FIXED TEXT — — 25¢

TELEGRAMS OF SENDER'S OWN COMPOSITION (first 15 words) — 35¢

CONGRATULATIONS ON
Anniversaries, Weddings
Birthdays, Commencement
Birth of a Child

HOTEL or TRANSPORTATION RESERVATIONS
TOURATE TELEGRAMS, for TRAVELERS (first 15 words) — 35¢

MISCELLANEOUS
Bon Voyage telegrams, "Fyi" telegrams
Kiddiegrams (No 50¢ rate), "Thank You" telegrams

ASK ANY WESTERN UNION OFFICE OR AGENCY FOR FULL INFORMATION

The body is 11 feet long and will carry 1,000 feet of 2½ inch double jacket hose. 200 gallon booster tank. Booster tanks up to 500 gallon capacity. L shape, extending full length and width of body, may be safely carried leaving space for an ample supply of hose. A 12 foot roof ladder and a 28 foot extension ladder are carried under the slatted floor. Also 3 lengths of hard suction hose. Ladders and hose held secure by single, clamping lock device, instantly released by one lever. Flexibility of design permits carriage of longer or additional ladders, suction hose, pony suction hose, and any tools.

The Cab comfortably seats two forward and six in back. Four wide doors for speedy entry and exit. Roll-up windows with safety glass all-around. Spring cushioned seats and back rests. Removable seats. Heavily upholstered and covered with specially fabricated waterproof and weatherproof material, more durable than leather.

will never be deflected and sag in middle or wear. Impeller is held in perfect alignment because no shaft deflection can occur.

PHOSPHOR BRONZE IMPELLER—Enclosed type impeller of phosphor bronze with single seal ring for higher efficiency, longer life, easier maintenance. All other pumps have two seal rings and much greater wearing surface.

BALANCED IMPELLER—Eliminates the possibility of a breakdown from end thrust on bearings. Only engineers of long experience realize the terrific tons of pressure on the impeller and casing when a fire pump operates at high pressure.

In Champion design the impeller thrust is BALANCED—an exclusive feature not in any other centrifugal pump. No danger of a cracked ball bearing; the very pressure of the water within the pump is diverted into slots so that end thrust is practically eliminated.

PACKING PROTECTED FROM ABRASIVES — By means of

BOOSTER INLET SHUTOFF VALVE—For shutting off the line from tank when pumping from suction or hydrant.

BOOSTER DISCHARGE SHUTOFF VALVE—For shutting off the line to reel when booster hose is not in use.

BRASS DRAIN COCK—Drains pump absolutely dry and provides full protection against freezing in cold weather.

SUCTION INLETS—Size 4 inch on the M500 Pump. Size 3½ inch on the M400 Pump. Built-in strainers for positive protection when pumping from hydrants.

GAUGES—Four, special heavy duty type, for Fire Apparatus. Two pressure gauges calibrated 0-300 lbs. Two compound gauges calibrated 0-30 inches and 0-300 lbs.

THE ONLY PUMP DESIGNED FOR COMMERCIAL CHASSIS

Charge to the account of _____ $ _____

WESTERN UNION 1206-B

R. B. WHITE PRESIDENT   NEWCOMB CARLTON CHAIRMAN OF THE BOARD   J. C. WILLEVER FIRST VICE-PRESIDENT

Send the following message, subject to the terms on back hereof, which are hereby agreed to

☞ Write your Order on this Blank for anything in Our Catalog.

To W. S. DARLEY & CO.
CHICAGO, ILL.

SEND THIS MESSAGE
WE GUARANTEE COLLECT
CHARGES PAID ON DELIVERY

☞ SEND THIS MESSAGE BY **TELEGRAPH** TO W. S. DARLEY & CO.
WE PAY ALL THE CHARGES WHEN MESSAGE IS RECEIVED. YOU ARE WELCOME TO THIS SERVICE.

You Can Install a Pump On Any Chassis In a Few Hours

versal joints exactly the same as the two now used connecting the short drive shaft to the transmission and propeller shaft of the chassis. After removing this drive shaft (only a wrench is needed for the job) the pump slips into position; having been engineered for your truck there is no trick or trouble to get the proper location —it just has to rest in the right position. Make your universal connections at both front and rear and the job is done except for drilling two holes (one on each side of frame) through which bolts are passed for anchoring the pump in position.

Our low price includes everything as described and illustrated. Everything needed to make a complete installation. With each pump we send Easy Blue Prints and Easy Instructions for mounting.

Why Pay $1000 or More? This Pump Is Underwriter Approved and Costs Only $645 or $625⁶⁵ Cash with Order or C.O.D.

66

Designed for all makes and models of cab-over-engine chassis. The Apparatus pictured here is built on a Ford 157 inch w. b. chassis with 95 Horsepower Mercury engine. Dual rear wheels. Any type of Champion Pump may be selected.

LARGE COMPARTMENTS

The truck shown is built with five compartments, two in each side and one in rear. Many more, amply large, with side or top openings, can be conveniently located. Wide, single or double doors and quick action safety spring locks. Protects all equipment from weather and dust.

Salvage covers, coats, boots, extinguishers, nozzles, siamese, adapters, axes, crow bars, first aid kits, resuscitators, etc., are easily accessible yet out of sight, nothing exposed to mar the sleek, streamlined contour of design.

Flexible in Design So That We Can Build to Your Specifications and Give You Any Construction Features

The body is 11 feet long and will carry 1,000 feet of 2½ inch double jacket hose. 200 gallon booster tank. Booster tanks up to 500 gallon capacity. L shape, extending full length and width of body, may be safely carried leaving space for an ample supply of hose. A 12 foot roof ladder and a 28 foot extension ladder are carried under the slatted floor. Also 3 lengths of hard suction hose. Ladders and hose held secure by single, clamping lock device, instantly released by one lever. Flexibility of design permits carriage of longer or additional ladders, suction hose, pony suction hose, and any tools.

The Cab comfortably seats two forward and six in back. Four wide doors for speedy entry and exit. Roll-up windows with safety glass all-around. Spring cushioned seats and back rests. Removable seats. Heavily upholstered and covered with specially fabricated waterproof and weatherproof material, more durable than leather.

NEW!

DARLEY MAGNETIC DIPPING NEEDLE WITH TELESCOPE HANDLE

Adjustable To Your Height
For Maximum Efficiency and Comfort

A new improved type of the instrument which has served Supts. so well for over a quarter century.

The nonmagnetic, chromium plated handle is in three sections which telescope or slide into each other. No matter whether you're tall or short or medium you can adjust the handle so that you can stand upright in comfort while moving the instrument over the ground. Without stooping you hold the Magnetic Dipping Needle just above the ground for maximum efficiency and sensitiveness.

As you approach the box and the needle starts to dip it's easy to hold the instrument steady with the new handle. Top knob prevents instrument from slipping out of your fingers.

No need to guess any more where service boxes are buried! With our wonderful Magnetic Dipping Needle you can spot them every time—in a few minutes—without any preliminary digging.

This is not a toy—nor a "fake"—nor an experiment. It has been tested and approved by MORE THAN 9,000 PRACTICAL, HARD-HEADED WATER AND GAS MEN WHO SAY IT'S THE BEST INSTRUMENT OF ITS KIND THEY EVER SAW.

These 9,000 Magnetic Dipping Needles have saved hundreds of thousands of dollars for their owners. They have saved days of delay—immeasurable damage from broken mains—and have won for themselves innumerable friends among every man who ever used one.

It is simple to use. There are no wires—no battery—no hocus pocus about this needle. It is merely a precision-made, full jeweled magnetic dip needle as carefully manufactured as the finest watch—and produced with the greatest care to insure accuracy.

Move one of these needles over the place where you think the buried shut off or service box is located, and if it dips, the box is there! No dip—no box. This we absolutely guarantee. We further guarantee that there is no finer dipping needle in the world than our Tungsten Magnetic—none more sensitive or more accurate.

Quit being a guesser! Quit wasting the time of your men—your own time—your company's money—your efforts! Quit ruining your disposition! Try out one of these dipping needles and if it does not do all and more than we claim, send it back. What offer could be fairer than this?

Because made from genuine Tungsten Steel, our needles take over 20% more magnetism than any other steel—hold it longer—and enable it to give 100% better results.

The jeweled bearings of our needles are as hard as a diamond, and as finely polished. They are genuine sapphires. The pivots in which the needles are mounted are steel of glass-like hardness. It takes over 120 operations to make this needle—it contains over 30 delicate parts—and countless inspections are required to insure perfection in manufacture and operation.

ALL THE YEAR
it is useful, as it works through ice and snow in winter and through earth, concrete, cement, macadam, asphalt, wood, brick pavements, sidewalks, cinders, etc.

$19.75

No. E999. Genuine Sapphire Jeweled Bearings. Chrome Tungsten Steel Needle. Chromium plated all over. New 3-section, telescope handle, adjustable to any length, chromium plated. 9,000 and more in use giving perfect satisfaction.

Price, complete, in velvet lined case, fully jeweled and carefully tested before shipment, with complete instructions, shipping charges prepaid...........$19.75

You May Try It Out On Suspicion
Test It Out a Full Month

It's easy to prove every claim made for the Magnetic Dipping Needle without risk and with nothing to pay until you're satisfied. Write a letter or send a post card asking us to ship on approval. Don't send any money. Try the Magnetic Dipping Needle thirty days before you decide whether it is worth our low price. Compute the savings it will mean. Satisfy yourself. This offer applies to both E999 and H605 Dipping Needles.

"Old Reliable"

Standard With Thousands of Water Works Men For Over 30 Years

Now Reduced To Only **$15**

The same highly efficient, super sensitive, dependable instrument, so popular with Supts. all over the world.

Quality the same, only the price changes.

No. H605. Complete in case, only......$15.00

TRADE IN YOUR OLD DIPPING NEEDLE

$5 ALLOWANCE

We Will Allow You $5 on the Purchase of the New No. E 999 Dipping Needle